UNDERSTANDING GERM WARFARE

UNDERSTANDING GERM WARFARE

FROM THE EDITORS OF *SCIENTIFIC AMERICAN*

Compiled and with introductions by

Sandy Fritz

Foreword by Dr. Jack Brown

A Byron Preiss Book

WARNER BOOKS

An AOL Time Warner Company

Copyright © 2002 by Scientific American, Inc., and Byron Preiss Visual Publications, Inc. All rights reserved.

The essays in this book first appeared in the pages and on the Web site of *Scientific American* as follows: "Evolution and the Origins of Disease" December 1996; "A Host with Infectious Ideas" May 2001; "The Challenge of Antibiotic Resistance" March 1998; "Fighting Bacterial Infections with Bacteria" January 2001; "The Cutting Edge" May 2001; "New Technique Resensitizes Resistant Bacteria to 'Last Resort' Antibiotic" August 2001; "Beyond Chicken Soup" November 2001; "Muscular Again" September 1999; "Many Ways to Make a Vaccine" January 1995; "Is Global Warming Harmful to Health?" August 2000; "Outbreak Not Contained" April 2000; "Trailing a Virus" August 1999; "Facing an Ill Wind" April 1999; "Bioagent Chip" March 2000; "The Specter of Biological Weapons" December 1996.

Illustration © 2002 Library of Congress: viii, x, 8, 13, 54, 64, 73. © 2002 Dennis Kunkel Microscopy, Inc: 125

Warner Books, Inc., 1271 Avenue of the Americas, New York, NY 10020
Visit our Web site at www.twbookmark.com.

 An AOL Time Warner Company

Printed in the United States of America
First Printing: December 2002
10 9 8 7 6 5 4 3 2

Library of Congress Cataloging-in-Publication Data

Understanding germ warfare / from the editors of Scientific American ; compiled and with introductions by Sandy Fritz ; foreword by Jack Brown.
   p. cm.
"A Byron Preiss book."
Includes index.
ISBN 0-446-67954-2
   1. Medical microbiology. 2. Microbial ecology. 3. Pathogenic microorganisms. 4. Biological weapons. 5. Drug resistance in microorganisms. I. Fritz, Sandy. II. Scientific American.

QR46 .U59 2002
616'.01—dc21
                                                 2002025895

Cover design by J. Vita
Book design by Gilda Hannah

Contents

lling bacteria is like hitt
rget--just when you think y
em, they mutate on you. Thr
eruse and misuse of antibio
d farmers have unwittingly
celerated the development o
cteria that survive convent
ic attack. To keep drugs po
e medical community preach
antibiotics and teach pati

Foreword

Dr. Jack Brown

F ossil records show that germs have inhabited the earth for a very long time—perhaps billions of years. And since the dawn of their history, human beings have been in almost constant conflict with germs.

Writings and images from thousands of years ago reveal early knowledge of this fight against germs. Paintings from the time of the Pharaohs show individuals clearly stricken with the disease now known as polio. In approximately 400 B.C., Hippocrates, the father of modern medicine, encouraged sanitation, or the killing of germs, as a disease preventative. Thucydides, a Greek historian from the same time, described how survivors of what was probably the plague were somehow protected from subsequent outbreaks of the same disease and could be used to assist the newly ill. A later outbreak of the plague, also known as the Black Death, began in Europe in the 1300s and cycled among European inhabitants for over 300 years. By the 1600s this disease alone was responsible for the loss of nearly 75% of the population of Europe. Although the plague abated due to more widespread application of sanitation, many other infectious diseases continued to claim the lives of individuals, particularly the young.

As recently as 70 to 80 years ago, a child had an estimated fifty-fifty chance of dying from a contracted infectious disease before the age of ten. Whooping cough, cholera, measles and other infectious diseases killed thousands of children in the United States each year. In some underdeveloped nations, this high probability of death from diseases continues to exist. In the 1940s, the widespread availability of antibiotics, particularly

Plague victim in 1500 pointing out boil to three physicians

penicillin, enabled great strides to be made to protect individuals from infections.

The development of vaccines in the 1950s and 1960s against many different bacteria, bacterial toxins and viruses led to dramatic success in preventing infectious illnesses. For example, in 1921 in the United States there were more than 100,000 reported cases of diphtheria. In 1996 there were fewer than 10 such cases—the dramatic decline the result a potent vaccine used against this bacterium. In 1941 there were approximately one million reported cases of the measles, but by 1996 vaccines led to less than 1,000 reported cases. In 1952 there were approximately 30,000 cases of polio reported in the United States, a number reduced by medical efforts to zero by 1996. However, the availability of vaccines and other armaments does not necessarily guarantee safety from infectious illness. It is estimated that in the United States alone at least 20,000 deaths occur each year due to complications associated with contracting the flu.

Each scientific step forward has led to new questions. How have the inexorable pressures of evolution shaped the organisms that cause disease and our response to them? Why do many antibiotics no longer work against certain bacteria? With all of the new technology, why aren't we further ahead in this fight? How can these germs be used against us in combat? Are we ready for any resurgence of these germs? These questions among others are thoroughly considered in this timely, well written and provocative collection of essays that focus upon germs and our relationship to them.

Understanding Germ Warfare begins with a remarkable journey into the complex world of germs through discussions of evolutionary biology, a field that is growing in importance by providing insight into the causative mechanisms that establish interdependence between germs and humans as each evolves. By understanding such relationships, the reader gains a better understanding of infectious disease and how humans may

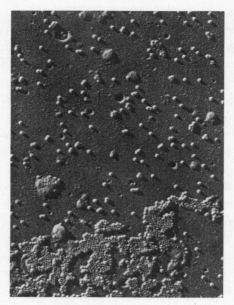

The first visualization of polio virus

impose restrictions on disease outcomes. Subsequent chapters focus upon various aspects of infectious disease and the methods used, including antibiotics and vaccines, to fight the bacteria and viruses that cause such harm.

While antibiotics have saved many lives, the overuse of these drugs in medicine, agriculture and animal care has also resulted in a selection of harmful bacteria that are now resistant. How do infectious organisms gain resistance to such drugs? How can we avoid or at least diminish this apparently inevitable outcome from the use of antibiotics? What new methodologies are presently being developed to combat germs? Can we significantly improve our own fighting response, the body's immune response, to infectious organisms, and if so, how? These issues are thoroughly discussed and potential solutions revealed in the upcoming chapters.

Understanding Germ Warfare also details how widespread travel affects global exposure to previously geographically isolated infectious agents; what affect global warming has upon infectious disease occurrences; and how rapidly and how well we can respond to the occurrence of a previously unknown infectious organism.

The last section of the book concerns biological warfare and related issues—a most complicated subject, particularly in light of the events since September 11, 2001. Bioterrorism, or the purposeful use of biologic agents to wreak death and fear among innocent populations, is not a new idea but sadly is currently a significant concern. These chapters discuss in a straightforward and non-inflammatory manner the biologic agents involved and offer timely suggestions as to the methods, both practical and political, that may be used to prevent and/or medically respond to such attacks.

When you have finished reading this collection, you will be brought up to speed on our on-going war with germs by authorities in almost every aspect of infectious organisms. *Understanding Germ Warfare* will enable you to travel through this complex micro-world on sure footing.

Introduction

Sandy Fritz

Germs are everywhere!" warns a soap advertisement. "Protect yourself and your family!" While it is true that germs are everywhere, antibiotic hand soap does little to protect you from the kind of germs that would cause harm.

Germs, a word derived from the Latin word meaning to grow, is a blanket term that describes microorganisms—mostly bacteria and viruses. The majority of these microbes live blameless lives, with many that actually help us. A few, however, have specialized in infesting human beings. Germs that infect us and make us sick are known as pathogens.

How did pathogens evolve? And how did the human body develop a defensive strategy to deal with an attack by life forms too small to be seen by the unaided eye? The delicate balance of defense against invaders while keeping the immune system calm enough to leave healthy body tissue undamaged is the topic of this book's first article, "Evolution and the Origin of Disease." In it are described instances where our defenses are nothing short of elegant. In other instances you'll learn our defenses resemble patched together solutions that are limited due to the very nature of our physical structure.

When the body itself has trouble handling infectious bacteria, antibiotics can help. Antibiotic drugs have only been around since the 1940s and, when discovered, were hailed as "miracle drugs." Unfortunately, misuse and overuse of antibiotics are giving rise to bacteria that resist treatment. More alarmingly, bacteria frequently freely exchange genetic material, so one type of bacteria that has developed resistance to antibiotic treatment could, in some cases, confer the resistance to another bacteria. These so-called "super germs" are taxing scientific brain power to develop new solutions to fight them.

Viruses, on the other hand, tend to be difficult to treat with drug therapy. The best defense against viral infection is vaccines, usually small amounts of a crippled viral pathogen that is deliberately injected into the human body. There, the immune system has a chance to "learn" the identity of the virus so that it can swiftly react to an attack at a future time. New technologies that read the genetic sequences of viral pathogens could lead to innovative new treatments. Still, technology itself—especially technology that enhances global mobility—can be a vector for opportunistic pathogens.

The ever-increasing globalization of our societies has, in some ways, opened up the proverbial Pandora's Box of pathogenic problems. Sometimes, increasing human population in the developing world awakens deadly, little-known pathogens by disturbing their ancient dwelling places. AIDS, an unknown and provincial virus just 25 years ago, is now a global threat. "Trailing a Virus," the story of what happens when a lethal virus new to science suddenly appears, underscores just how at risk the entire human family can be.

Throughout history, countless millions of people have been waylaid and slain by deadly pathogens. The cycles of rising and falling epidemics are frequently tuned to natural environmental disasters, such as floods or the movements of populations from one area of the world to another. In "Is Global Warming Hazardous to Your Health," we get a sneak preview of some of the dynamics that could trigger worldwide health problems. As terrible as such outbreaks are, they are at least couched in nature, and tend to wane as the environment stabilizes again.

Curiously, as scientists come to grips with the nature of pathogens, they have created a whole new set of probabilities for death and danger. Over the past 150 years, scientist have learned to identify, isolate, and grow pathogens in the laboratory, ostensibly for the purpose of studying them to better find a cure. Now, the ability to grow deadly pathogens has taken the perverse path of warfare. It is possible for a college-educated biology graduate to cull, grow, and develop weapons of war that in many ways are far more terrible than so-called conventional weapons. Leonard Cole explores this subject in depth in his essay "The Specter of Biological Weapons."

As our understanding and abilities grow, the need for responsible action grows too. Some germs are killers. And they can be used as agents of assassination not only against individuals, but against whole populations. Our world stands close to the brink of either breakthrough or breakdown. The accumulated wisdom and insight of our sciences can either help limit the spread of disease and misery, or it can promote global disease and misery. The choice is our own.

It is a fact of life: people get sick. A stomach bug may overwhelm us. Or an influenza attack triggers the flu. Or chicken pox raises itchy welts.

Why we get sick and how our bodies respond to infectious pathogens in certain ways is determined by our evolution. Complex and elegant natural defenses, shaped over a vast span of time, guard the body from a bevy of well-known illness. Yet, these mechanisms are flexible enough to innovate as new diseases appear. The following two articles detail how and why our bodies evolve.

Evolution and the Origins of Disease

Randolph M. Nesse and George C. Williams

Thoughtful contemplation of the human body elicits awe—in equal measure with perplexity. The eye, for instance, has long been an object of wonder, with the clear, living tissue of the cornea curving just the right amount, the iris adjusting to brightness and the lens to distance, so that the optimal quantity of light focuses exactly on the surface of the retina. Admiration of such apparent perfection soon gives way, however, to consternation. Contrary to any sensible design, blood vessels and nerves traverse the inside of the retina, creating a blind spot at their point of exit.

The body is a bundle of such jarring contradictions. For each exquisite heart valve, we have a wisdom tooth. Strands of DNA direct the development of the 10 trillion cells that make up a human adult but then permit his or her steady deterioration and eventual death. Our immune system can identify and destroy a million kinds of foreign matter, yet many bacteria can still kill us. These contradictions make it appear as if the body was designed by a team of superb engineers with occasional interventions by Rube Goldberg.

In fact, such seeming incongruities make sense but only when we investigate the origins of the body's vulnerabilities while keeping in mind the wise words of distinguished geneticist Theodosius Dobzhansky: "Nothing in biology makes sense except in the light of evolution." Evolutionary biology is, of course, the scientific foundation for all biology, and biology is the foundation for all medicine. To a surprising degree, however, evolutionary biology is just now being recognized as a basic medical science. The enterprise of studying medical problems in an evolutionary context has been termed Darwinian medicine. Most medical research tries to explain the causes of an individual's disease and seeks therapies to cure or relieve deleterious conditions. These efforts are traditionally based on consideration of proximate issues, the straightforward study of the body's anatomic and physiological mechanisms as they currently exist. In contrast, Darwinian medicine asks why the body is designed in a way that makes us all vulnerable to problems like cancer, atherosclerosis, depression and choking, thus offering a broader context in which to conduct research.

The evolutionary explanations for the body's flaws fall into surprisingly few categories. First, some discomforting conditions, such as pain, fever, cough, vomiting and anxiety, are actually neither diseases nor design defects but rather are evolved defenses. Second, conflicts with other organisms—*Escherichia coli* or crocodiles, for instance—are a fact of life. Third, some circumstances, such as the ready availability of dietary fats, are so recent that natural selection has not yet had a chance to deal with them. Fourth, the body may fall victim to trade-offs between a trait's benefits and its costs; a textbook example is the sickle cell gene, which also protects against malaria. Finally, the process of natural selection is constrained in ways that leave us with suboptimal design features, as in the case of the mammalian eye.

Perhaps the most obviously useful defense mechanism is coughing; people who cannot clear foreign matter from their lungs are likely to die from pneumonia. The capacity for pain is also certainly beneficial. The rare individuals who cannot feel pain fail even to experience discomfort from staying in the same position for long periods. Their unnatural stillness impairs the blood supply to their joints, which then deteriorate. Such pain-free people usually die by early adulthood from tissue damage and infections. Cough or pain is usually interpreted as disease or trauma but is actually part of the solution rather than the problem. These defensive capabilities, shaped by natural selection, are kept in reserve until needed.

Less widely recognized as defenses are fever, nausea, vomiting, diarrhea, anxiety, fatigue, sneezing and inflammation. Even some physicians remain unaware of fever's utility. No mere increase in metabolic rate, fever is a carefully regulated rise in the set point of the body's thermostat. The higher body temperature facilitates the destruction of pathogens. Work by Matthew J. Kluger has shown that even cold-blooded lizards, when infected, move to warmer places until their bodies are several degrees above their usual temperature. If prevented from moving to the warm part of their cage, they are at increased risk of death from the infection. In a similar study by Evelyn Satinoff elderly rats, who can no longer achieve the high fevers of their younger lab companions, also instinctively sought hotter environments when challenged by infection.

A reduced level of iron in the blood is another misunderstood defense mechanism. People suffering from chronic infection often have decreased levels of blood iron. Although such low iron is sometimes blamed for the illness, it actually is a protective response: during infection, iron is sequestered in the liver, which prevents invading bacteria from getting adequate supplies of this vital element.

Morning sickness has long been considered an unfortunate side effect of pregnancy. The nausea, however, coincides with the period of rapid tissue differentiation of the fetus, when development is most vulnerable to interference by toxins. And

"Make America Strong" poster from World War II

nauseated women tend to restrict their intake of strong-tasting, potentially harmful substances. These observations led independent researcher Margie Profet to hypothesize that the nausea of pregnancy is an adaptation whereby the mother protects the fetus from exposure to toxins. Profet tested this idea by examining pregnancy outcomes. Sure enough, women with more nausea were less likely to suffer miscarriages. (This evidence supports the hypothesis but is hardly conclusive. If Profet is correct, further research should discover that pregnant females of many species show changes in food preferences. Her theory also predicts an increase in birth defects among offspring of women who have little or no morning sickness and thus eat a wider variety of foods during pregnancy.)

Another common condition, anxiety, obviously originated as a defense in dangerous situations by promoting escape and avoidance. A 1992 study by Lee A. Dugatkin evaluated the benefits of fear in guppies. He grouped them as timid, ordinary or bold, depending on their reaction to the presence of small-mouth bass. The timid hid, the ordinary simply swam away, and the bold maintained their ground and eyed the bass. Each guppy group was then left alone in a tank with a bass. After 60 hours, 40 percent of the timid guppies had survived, as had only 15 percent of the ordinary fish. The entire complement of bold guppies, on the other hand, wound up aiding the transmission of bass genes rather than their own.

Selection for genes promoting anxious behaviors implies that there should be people who experience too much anxiety, and indeed there are. There should also be hypophobic individuals who have insufficient anxiety, either because of genetic tendencies or antianxiety drugs. The exact nature and frequency of such a syndrome is an open question, as few people come to psychiatrists complaining of insufficient apprehension. But if sought, the pathologically nonanxious may be found in emergency rooms, jails and unemployment lines.

The utility of common and unpleasant conditions such as diarrhea, fever and anxiety is not intuitive. If natural selection shapes the mechanisms that regulate defensive responses, how can people get away with using drugs to block these defenses without doing their bodies obvious harm? Part of the answer is that we do, in fact, sometimes do ourselves a disservice by disrupting defenses.

Herbert L. DuPont and Richard B. Hornick studied the

diarrhea caused by Shigella infection and found that people who took antidiarrhea drugs stayed sick longer and were more likely to have complications than those who took a placebo. In another example, Eugene D. Weinberg has documented that well-intentioned attempts to correct perceived iron deficiencies have led to increases in infectious disease, especially amebiasis, in parts of Africa. Although the iron in most oral supplements is unlikely to make much difference in otherwise healthy people with everyday infections, it can severely harm those who are infected and malnourished. Such people cannot make enough protein to bind the iron, leaving it free for use by infectious agents.

On the morning-sickness front, an antinausea drug was recently blamed for birth defects. It appears that no consideration was given to the possibility that the drug itself might be harmless to the fetus but could still be associated with birth defects, by interfering with the mother's defensive nausea.

Another obstacle to perceiving the benefits of defenses arises from the observation that many individuals regularly experience seemingly worthless reactions of anxiety, pain, fever, diarrhea or nausea. The explanation requires an analysis of the regulation of defensive responses in terms of signal-detection theory. A circulating toxin may come from something in the stomach. An organism can expel it by vomiting, but only at a price. The cost of a false alarm—vomiting when no toxin is truly present—is only a few calories. But the penalty for a single missed authentic alarm—failure to vomit when confronted with a toxin—may be death.

Natural selection therefore tends to shape regulation mechanisms with hair triggers, following what we call the smoke-detector principle. A smoke alarm that will reliably wake a sleeping family in the event of any fire will necessarily give a false alarm every time the toast burns. The price of the human body's numerous "smoke alarms" is much suffering that is completely normal but in most instances unnecessary. This principle also explains why blocking defenses is so often free of tragic consequences. Because most defensive reactions occur in response to insignificant threats, interference is usually harmless; the vast majority of alarms that are stopped by removing the battery from the smoke alarm are false ones, so this strategy may seem reasonable. Until, that is, a real fire occurs.

Natural selection is unable to provide us with perfect protection against all pathogens, because they tend to evolve much faster than humans do. *Escherichia coli*, for example, with its rapid rates of reproduction, has as much opportunity for mutation and selection in one day as humanity gets in a millennium. And our defenses, whether natural or artificial, make for potent selection forces. Pathogens either quickly evolve a counterdefense or become extinct. Biologist Paul W. Ewald has suggested classifying phenomena associated with infection according to whether they benefit the host, the pathogen, both or neither. Consider the runny nose associated with a cold. Nasal mucous secretion could expel intruders, speed the pathogen's transmission to new hosts or both. Answers could come from studies examining whether blocking nasal secretions shortens or prolongs illness, but few such studies have been done.

Humanity won huge battles in the war against pathogens with the development of antibiotics and vaccines. Our victories were so rapid and seemingly complete that in 1969 U.S. Surgeon General William H. Stewart said that it was "time to close the book on infectious disease." But the enemy, and the power of natural selection, had been underestimated. The sober reality is that pathogens apparently can adapt to every chemical researchers develop. ("The war has been won," one scientist more recently quipped. "By the other side.")

Antibiotic resistance is a classic demonstration of natural selection. Bacteria that happen to have genes that allow them to prosper despite the presence of an antibiotic reproduce faster than others, and so the genes that confer resistance spread quickly. As shown by Nobel laureate Joshua Lederberg, they can even jump to different species of bacteria, borne on bits of infectious DNA. Today some strains of tuberculosis in New York City are resistant to all three main antibiotic treatments; patients with those strains have no better chance of surviving than did TB patients a century ago. Stephen S. Morse notes that the multidrug-resistant strain that has spread throughout the East Coast may have originated in a homeless shelter across the street from Columbia-Presbyterian Medical Center. Such a phenomenon would indeed be predicted in an environment where fierce selection pressure quickly weeds out less hardy strains. The surviving bacilli have been bred for resistance (see "The Challenge of Antibiotic Resistance," page 27).

Many people, including some physicians and scientists, still believe the theory that pathogens necessarily become benign after long association with hosts. Superficially, this makes sense. An organism that kills rapidly may never get to a new host, so natural selection would seem to favor lower virulence. Syphilis, for instance, was a highly virulent disease when it first arrived in Europe, but as the centuries passed it became steadily more mild. The virulence of a pathogen is, however, a life history trait that can increase as well as decrease, depending on which option is more advantageous to its genes (see "A Host with Infectious Ideas," page 21).

For agents of disease that are spread directly from person to person, low virulence tends to be beneficial, as it allows the host to remain active and in contact with other potential hosts. But some diseases, like malaria, are transmitted just as well— or better—by the incapacitated. For such pathogens, which usually rely on intermediate vectors like mosquitoes, high virulence can give a selective advantage. This principle has direct implications for infection control in hospitals, where health care workers' hands can be vectors that lead to selection for more virulent strains.

1941 poster warning parents about methods of transmission

In the case of cholera, public water supplies play the mosquitoes' role. When water for drinking and bathing is contaminated by waste from immobilized patients, selection tends to increase virulence, because more diarrhea enhances the spread of the organism even if individual hosts quickly die. But, as Ewald has shown, when sanitation improves, selection acts against classical *Vibrio cholerae* bacteria in favor of the more benign El Tor biotype. Under these conditions, a dead host is a dead end. But a less ill and more mobile host, able to infect many others over a much longer time, is an effective vehicle for a pathogen of lower virulence. In another example, better sanitation leads to displacement of the aggressive *Shigella flexneri* by the more benign S. *sonnei*.

Such considerations may be relevant for public policy. Evolutionary theory predicts that clean needles and the encouragement of safe sex will do more than save numerous individuals from HIV infection. If humanity's behavior itself slows HIV transmission rates, strains that do not soon kill their hosts have the long-term survival advantage over the more virulent viruses that then die with their hosts, denied the opportunity to spread. Our collective choices can change the very nature of HIV.

Making rounds in any modern hospital provides sad testimony to the prevalence of diseases humanity has brought on itself. Heart attacks, for example, result mainly from atherosclerosis, a problem that became widespread only in this century and that remains rare among hunter-gatherers. Epidemiological research furnishes the information that should help us prevent heart attacks: limit fat intake, eat lots of vegetables, and exercise hard each day. But hamburger chains proliferate, diet foods languish on the shelves, and exercise machines serve as expensive clothing hangers throughout the land. The proportion of overweight Americans is one third and rising. We all know what is good for us. Why do so many of us continue to make unhealthy choices?

Our poor decisions about diet and exercise are made by brains shaped to cope with an environment substantially different from the one our species now inhabits. On the African savanna, where many believe the modern human design was fine-tuned, fat, salt and sugar were scarce and precious. Individuals who had a tendency to consume large amounts of fat when given the rare opportunity had a selective advantage. They were more likely to survive famines that killed their thinner companions. And we, their descendants, still carry those urges for foodstuffs that today are anything but scarce. These evolved desires—inflamed by advertisements from competing food corporations that themselves survive by selling us more of whatever we want to buy—easily defeat our intellect and willpower. How ironic that humanity worked for centuries to create environments that are almost literally flowing with milk and honey, only to see our success responsible for much modern disease and untimely death.

Increasingly, people also have easy access to many kinds of drugs, especially alcohol and tobacco, that are responsible for a huge proportion of disease, health care costs and premature death. Although individuals have always used psychoactive substances, widespread problems materialized only following another environmental novelty: the ready availability of concentrated drugs and new, direct routes of administration, especially injection. Most of these substances, including nicotine, cocaine and opium, are products of natural selection that evolved to protect plants from insects. Because humans share a common evolutionary heritage with insects, many of these substances also affect our nervous system.

This perspective suggests that it is not just defective individuals or disordered societies that are vulnerable to the dangers of psychoactive drugs; all of us are susceptible because drugs and our biochemistry have a long history of interaction. Understanding the details of that interaction, which is the focus of much current research from both a proximate and evolutionary perspective, may well lead to better treatments for addiction.

Compromise is inherent in every adaptation. Arm bones

three times their current thickness would almost never break, but Homo sapiens would be lumbering creatures on a never-ending quest for calcium. More sensitive ears might sometimes be useful, but we would be distracted by the noise of air molecules banging into our eardrums.

Such trade-offs also exist at the genetic level. If a mutation offers a net reproductive advantage, it will tend to increase in frequency in a population even if it causes vulnerability to disease. People with two copies of the sickle cell gene, for example, suffer terrible pain and die young. People with two copies of the "normal" gene are at high risk of death from malaria. But individuals with one of each are protected from both malaria and sickle cell disease. Where malaria is prevalent, such people are fitter, in the Darwinian sense, than members of either other group. So even though the sickle cell gene causes disease, it is selected for where malaria persists. Which is the "healthy" allele in this environment? The question has no answer. There is no one normal human genome—there are only genes.

Many other genes that cause disease must also have offered benefits, at least in some environments, or they would not be so common. Because cystic fibrosis (CF) kills one out of 2,500 Caucasians, the responsible genes would appear to be at great risk of being eliminated from the gene pool. And yet they endure. For years, researchers mused that the CF gene, like the sickle cell gene, probably conferred some advantage. Recently a study by Gerald B. Pier and his colleagues gave substance to this informed speculation: having one copy of the CF gene appears to decrease the chances of the bearer acquiring a typhoid fever infection, which once had a 15 percent mortality.

Aging may be the ultimate example of a genetic trade-off. In 1957 one of us (Williams) suggested that genes that cause aging and eventual death could nonetheless be selected for if they had other effects that gave an advantage in youth, when the force of selection is stronger. For instance, a hypothetical gene that governs calcium metabolism so that bones heal quickly but that also happens to cause the steady deposition of calcium in arterial walls might well be selected for even though it kills some older people. The influence of such pleiotropic genes (those having multiple effects) has been seen in fruit flies and flour beetles, but no specific example has yet been found in humans. Gout, however, is of particular interest, because it arises when a potent antioxidant, uric acid, forms crystals that precipitate out of fluid in joints. Antioxidants have antiaging effects, and plasma levels of uric acid in different species of primates are closely correlated with average adult life span. Perhaps high levels of uric acid benefit most humans by slowing tissue aging, while a few pay the price with gout.

Other examples are more likely to contribute to more rapid aging. For instance, strong immune defenses protect us from infection but also inflict continuous, low-level tissue damage. It is also possible, of course, that most genes that cause aging have no benefit at any age—they simply never decreased reproductive fitness enough in the natural environment to be selected against. Nevertheless, over the next decade research will surely identify specific genes that accelerate senescence, and researchers will soon thereafter gain the means to interfere with their actions or even change them. Before we tinker, however, we should determine whether these actions have benefits early in life.

Because evolution can take place only in the direction of time's arrow, an organism's design is constrained by structures already in place. As noted, the vertebrate eye is arranged backward. The squid eye, in contrast, is free from this defect, with vessels and nerves running on the outside, penetrating where necessary and pinning down the retina so it cannot detach. The human eye's flaw results from simple bad luck; hundreds of millions of years ago, the layer of cells that happened to become sensitive to light in our ancestors was positioned differently from the corresponding layer in ancestors of squids.

The two designs evolved along separate tracks, and there is no going back.

Such path dependence also explains why the simple act of swallowing can be life-threatening. Our respiratory and food passages intersect because in an early lungfish ancestor the air opening for breathing at the surface was understandably located at the top of the snout and led into a common space shared by the food passageway. Because natural selection cannot start from scratch, humans are stuck with the possibility that food will clog the opening to our lungs.

The path of natural selection can even lead to a potentially fatal cul-de-sac, as in the case of the appendix, that vestige of a cavity that our ancestors employed in digestion. Because it no longer performs that function, and as it can kill when infected, the expectation might be that natural selection would have eliminated it. The reality is more complex. Appendicitis results when inflammation causes swelling, which compresses the artery supplying blood to the appendix. Blood flow protects against bacterial growth, so any reduction aids infection, which creates more swelling. If the blood supply is cut off completely, bacteria have free rein until the appendix bursts. A slender appendix is especially susceptible to this chain of events, so appendicitis may, paradoxically, apply the selective pressure that maintains a large appendix. Far from arguing that everything in the body is perfect, an evolutionary analysis reveals that we live with some very unfortunate legacies and that some vulnerabilities may even be actively maintained by the force of natural selection.

Despite the power of the Darwinian paradigm, evolutionary biology is just now being recognized as a basic science essential for medicine. Most diseases decrease fitness, so it would seem that natural selection could explain only health, not disease. A Darwinian approach makes sense only when the object of explanation is changed from diseases to the traits that make us vulnerable to diseases. The assumption that natural selection maximizes health also is incorrect—selection maximizes the reproductive success of genes. Those genes that make bodies having superior reproductive success will become more common, even if they compromise the individual's health in the end.

Finally, history and misunderstanding have presented obstacles to the acceptance of Darwinian medicine. An evolutionary approach to functional analysis can appear akin to naive teleology or vitalism, errors banished only recently, and with great effort, from medical thinking. And, of course, whenever evolution and medicine are mentioned together, the specter of eugenics—the deliberate attempt to "improve" the human species—arises. Discoveries made through a Darwinian view of how all human bodies are alike in their vulnerability to disease will offer great benefits for individuals, but such insights do not imply that we can or should make any attempt to improve the species. If anything, this approach cautions that apparent genetic defects may have unrecognized adaptive significance, that a single "normal" genome is nonexistent and that notions of "normality" tend to be simplistic.

The systematic application of evolutionary biology to medicine is a new enterprise. Like biochemistry at the beginning of this century, Darwinian medicine very likely will need to develop in several incubators before it can prove its power and utility. If it must progress only from the work of scholars without funding to gather data to test their ideas, it will take decades for the field to mature. Departments of evolutionary biology in medical schools would accelerate the process, but for the most part they do not yet exist. If funding agencies had review panels with evolutionary expertise, research would develop faster, but such panels remain to be created.

The evolutionary viewpoint provides a deep connection between the states of disease and normal functioning and can integrate disparate avenues of medical research as well as suggest fresh and important areas of inquiry. Its utility and power will ultimately lead to recognition of evolutionary biology as a basic medical science.

A Host with Infectious Ideas

Steve Mirsky

Newton had a falling apple. Darwin mused on finches. Paul W. Ewald's inspiration was diarrhea. "I wish I had something more romantic," says the Amherst College evolutionary biologist. It gets uglier: Ewald, then a graduate student studying bird behavior, was camped near a Kansas garbage dump. As he waged a three-day battle against his sea of troubles, he contemplated the interactions between a host—himself, in this case—and a pathogen. "There's some organism in there," Ewald remembers thinking during that 1977 experience, "and this diarrhea might be my way of getting rid of the organism—or it might be the organism's way of manipulating my body" to maximize its chances of passage to the next victim by, for example, contaminating the water supply. "If it's a manipulation and you treat it, you're avoiding damage," he notes. "But if it's a defense and you treat it, you sabotage the host."

Host-pathogen relationships have dominated Ewald's thoughts ever since, leading to numerous articles, two books and, depending on whom you talk to, the respect or scorn of scientists and physicians. The admiration comes from those who think he was on to something really big in his earlier publications, which he summed up in his 1994 book *Evolution of Infectious Disease.* "I think that Paul Ewald has been a pioneer in using evolutionary theory to attack hard questions in pathogenesis," comments Stephen Morse, a virologist and epidemiologist at Columbia University. "His work has, for the first time, shown a way to generate testable hypotheses to study such questions as the evolution of virulence—once thought intractable—and infectious causes of chronic diseases." Indeed, the *Atlantic Monthly* referred to Ewald as "the Darwin of the microworld" (to which Ewald responds, "No, Darwin is Darwin of the microworld, too").

Any antipathy is the result of his latest research, outlined in *Plague Time*, published in 2000. The 47-year-old Ewald argued in the book that infection may play a role in cancer, atherosclerosis, Alzheimer's and other chronic conditions ordinarily thought of as inevitable consequences of genetics, lifestyle or aging. "Some of his recent work is controversial," Morse states. "I'd personally prefer to reserve judgment for now on those questions, at least until more data are in." Others are less gracious. One prominent atherosclerosis researcher politely panned Ewald in public but privately referred to his ideas using an eight-letter word, the first half of which is "bull."

In an April 1993 *Scientific American* article, Ewald smashed the old, and unfortunately still widely accepted, notion that parasites and their hosts inevitably evolve toward a benign coexistence. The tendency toward benignity is reserved for conditions passed directly from person to person. Someone too sick to mingle with others would indeed be a dead end for the most dangerous infections, but Ewald showed that infectious agents that use intermediate vectors for transmission, such as malaria's mosquitoes and cholera's contaminated water, are free to evolve toward greater destructive power. After all, a mosquito is free to feed on the sickest malaria victims and thus pass on the worst pathogens. Even more provocative was Ewald's exegesis on our potential to drive the evolution of pathogens through judicious public health measures. "The evolutionary hypothesis says that if you can make it so that sick people cannot pass on infections and that only healthy people can, you should favor the evolution of more benign strains," he explains.

Ewald ponders the evolutionary interplay between microbes and large organisms such as ourselves. Ewald suggests an experiment that could never be ethically done: "Select two countries, one with bad water and one with clean water, and introduce cholera into both." Theory holds that water in which microbes can thrive serves as a vector that lets dangerous virulence continue or worsen. On the other hand, treated water would kill cholera strains relying on diarrhea for transport; only mild strains would survive because their hosts would be healthy enough to transmit the pathogen directly to other people. "Essentially, that's what happened in 1991," Ewald says, referring to a cholera outbreak in Peru that spread through Latin America. He and his students analyzed cholera from Peru and Guatemala, which has unsafe water, and from Chile, whose water is trustworthy. They found that over the 1990s Chile's cholera did indeed become less virulent, whereas highly toxic strains persisted in the other countries. This concept should motivate public health officials to do things they should already be doing anyway, such as providing safe water and mosquito-proof housing. Although these ideas have yet to permeate medical school curricula fully, they seem beyond reproach theoretically. When Ewald wanders into the fields of chronic disease, however, he steps into some eight-letter castigation. Given evolutionary principles and the available evidence, he argues in *Plague Time*, infectious agents should be considered as at least part of the etiology of apparently noninfectious conditions. Of course, the connection between *Helicobactor pylori* and peptic ulcers is now taken for granted, although medical texts of 20 years ago were mute on the subject. Associations between infections and some cancers—hepatitis virus with liver cancer, papillomavirus with cervical cancer—have become accepted in only the past few decades. Ewald thinks that more cancers, perhaps the majority, as well as numerous other common, widespread and ancient chronic diseases, will eventually become linked with various infections: for atherosclerosis and Alzheimer's disease, he points to studies showing associations with *Chlamydia pneumoniae*. He even holds that schizophrenia may be related to infection with the protozoan *Toxoplasma gondii*.

"People have put much more emphasis on genetic causation and noninfectious environmental causation," Ewald says. "And when they find evidence that those kinds of causation are occurring, then they make this fundamental error in science: throwing out a hypothesis [infection] just because you have evidence that other hypotheses are probably at least partly right." Disease instead may result from a subtle interplay between a gene's product and an infectious agent.

Arguably, natural selection should have gotten rid of most of the solely genetic diseases long ago. (Genetic conditions such as sickle-cell disease get an evolutionary pass, however: one copy of the gene protects against disease—malaria, in the case of sickle cell—so the potentially destructive gene will survive in a population.) The standard argument is that genes that cause illness after the prime reproductive years don't get selected against. Ewald counters by arguing that the elderly—and he believes that there were always people who would be considered old by today's standards, even at times when life was supposed to be "nasty, brutish and short"—were important sources of information and caregiving, and evolution does indeed try to keep them around.

To find possible infectious relationships to seemingly noninfectious diseases, Ewald suggests the creation of a program akin to that used to monitor adverse reactions to vaccines:

what he calls the Effects of Antimicrobials Reporting System, or EARS. Physicians worldwide may be sitting on a gold mine of data, in the form of anecdotes about remissions that accompany antibiotic treatment for a concurrent condition. "If you accumulate the shared experiences, real cause and effect should pop out," he says. "Then we'd know if this was something we should do a controlled study on."

Ewald believes that the associations between chronic diseases and infections will be slowly accepted, perhaps in a few decades. Should his viewpoint prevail some distant day, he may repeat the words his physicist father once spoke. The elder Ewald, recovering from a heart attack when Paul's 1993 article appeared in *Scientific American*, his favorite publication, said, "Well, this was worth living for."

Bacteria are living creatures. When their survival is threatened, they react. The continued attack on infectious bacteria puts pressure on the bacteria to evolve and change to survive. In some ways, our own medicine has given rise to "supergerms," recently emerging strains of bacteria that resist even the strongest treatments.

With the emergence of supergerms, microbiologists are on the hot seat to find new solutions and new treatments. A variety of tactics, outlined in the following articles, are in development that could lead to new drugs that really work.

lling bacteria is like hitt
rget--just when you think y
em, they mutate on you. Thr
eruse and misuse of antibio
d farmers have unwittingly
celerated the development o
cteria that survive convent
ic attack. To keep drugs po
e medical community preach
antibiotics and teach pati

The Challenge of Antibiotic Resistance

Stuart B. Levy

I n 1997 an event doctors had been fearing finally occurred. In three geographically separate patients, an often deadly bacterium, *Staphylococcus aureus*, responded poorly to a once reliable antidote—the antibiotic vancomycin. Fortunately, in those patients, the staph microbe remained susceptible to other drugs and was eradicated. But the appearance of *S. aureus* not readily cleared by vancomycin foreshadows trouble.

Worldwide, many strains of *S. aureus* are already resistant to all antibiotics except vancomycin. Emergence of forms lacking sensitivity to vancomycin signifies that variants untreatable by every known antibiotic are on their way. *S. aureus*, a major cause of hospital-acquired infections, has thus moved one step closer to becoming an unstoppable killer.

The looming threat of incurable *S. aureus* is just the latest twist in an international public health nightmare: increasing bacterial resistance to many antibiotics that once cured bacterial diseases readily. Ever since antibiotics became widely available in the 1940s, they have been hailed as miracle drugs—magic bullets able to eliminate bacteria without doing much harm to the cells of treated individuals. Yet with each passing decade, bacteria that defy not only single but multiple antibiotics—and therefore are extremely difficult to control—have become increasingly common.

What is more, strains of at least three bacterial species capable of causing life-threatening illnesses (*Enterococcus faecalis, Mycobacterium tuberculosis* and *Pseudomonas aeruginosa*) already evade every antibiotic in the clinician's armamentarium, a stockpile of more than 100 drugs. In part because of the rise in resistance to antibiotics, the death rates for some communicable diseases (such as tuberculosis) have started to rise again, after having declined in the industrial nations.

How did we end up in this worrisome, and worsening, situation? Several interacting processes are at fault. Analyses of them point to a number of actions that could help reverse the trend, if individuals, businesses and governments around the world can find the will to implement them.

One component of the solution is recognizing that bacteria are a natural, and needed, part of life. Bacteria, which are microscopic, single-cell entities, abound on inanimate surfaces and on parts of the body that make contact with the outer world, including the skin, the mucous membranes and the lining of the intestinal tract. Most live blamelessly. In fact, they often protect us from disease, because they compete with, and thus limit the proliferation of, pathogenic bacteria— the minority of species that can multiply aggressively (into the millions) and damage tissues or otherwise cause illness. The benign competitors can be important allies in the fight against antibiotic-resistant pathogens.

People should also realize that although antibiotics are needed to control bacterial infections, they can have broad, undesirable effects on microbial ecology. That is, they can produce long-lasting change in the kinds and proportions of bacteria—and the mix of antibiotic-resistant and antibiotic-susceptible types—not only in the treated individual but also in the environment and society at large. The compounds should thus be used only when they are truly needed, and they should not be administered for viral infections, over which they have no power.

Although many factors can influence whether bacteria in a person or in a community will become insensitive to an antibiotic, the two main forces are the prevalence of resistance genes (which give rise to proteins that shield bacteria from an antibiotic's effects) and the extent of antibiotic use. If the collective bacterial flora in a community have no genes conferring resistance to a given antibiotic, the antibiotic will successfully eliminate infection caused by any of the bacterial species in the collection. On the other hand, if the flora possess resistance genes and the community uses the drug persistently, bacteria able to defy eradication by the compound will emerge and multiply.

Antibiotic-resistant pathogens are not more virulent than susceptible ones: the same numbers of resistant and susceptible bacterial cells are required to produce disease. But the resistant forms are harder to destroy. Those that are slightly insensitive to an antibiotic can often be eliminated by using more of the drug; those that are highly resistant require other therapies.

To understand how resistance genes enable bacteria to survive an attack by an antibiotic, it helps to know exactly what antibiotics are and how they harm bacteria. Strictly speaking, the compounds are defined as natural substances (made by living organisms) that inhibit the growth, or proliferation, of bacteria or kill them directly. In practice, though, most commercial antibiotics have been chemically altered in the laboratory to improve their potency or to increase the range of species they affect. Here I will also use the term to encompass completely synthetic medicines, such as quinolones and sulfonamides, which technically fit under the broader rubric of antimicrobials.

Whatever their monikers, antibiotics, by inhibiting bacterial growth, give a host's immune defenses a chance to outflank the bugs that remain. The drugs typically retard bacterial proliferation by entering the microbes and interfering with the production of components needed to form new bacterial cells. For instance, the antibiotic tetracycline binds to ribosomes (internal structures that make new proteins) and, in so doing, impairs protein manufacture; penicillin and vancomycin impede proper synthesis of the bacterial cell wall.

Certain resistance genes ward off destruction by giving rise to enzymes that degrade antibiotics or that chemically modify, and so inactivate, the drugs. Alternatively, some resistance genes cause bacteria to alter or replace molecules that are normally bound by an antibiotic—changes that essentially eliminate the drug's targets in bacterial cells. Bacteria might also eliminate entry ports for the drugs or, more effectively, may manufacture pumps that export antibiotics before the medicines have a chance to find their intracellular targets.

Bacteria can acquire resistance genes through a few routes. Many inherit the genes from their forerunners. Other times, genetic mutations, which occur readily in bacteria, will spontaneously produce a new resistance trait or will strengthen an existing one. And frequently, bacteria will gain a defense against an antibiotic by taking up resistance genes from other bacterial cells in the vicinity. Indeed, the exchange of genes is so pervasive that the entire bacterial world can be thought of as one huge multicellular organism in which the cells interchange their genes with ease.

Bacteria have evolved several ways to share their resistance traits with one another. Resistance genes commonly are carried on plasmids, tiny loops of DNA that can help bacteria survive various hazards in the environment. But the genes may also occur on the bacterial chromosome, the larger DNA molecule that stores the genes needed for the reproduction and routine maintenance of a bacterial cell.

Often one bacterium will pass resistance traits to others by giving them a useful plasmid. Resistance genes can also be transferred by viruses that occasionally extract a gene from one bacterial cell and inject it into a different one. In addition, after a bacterium dies and releases its contents into the environment, another will occasionally take up a liberated gene for itself.

In the last two situations, the gene will survive and provide protection from an antibiotic only if integrated stably into a plasmid or chromosome. Such integration occurs frequently, though, because resistance genes are often embedded in small units of DNA, called transposons, that readily hop into other DNA molecules. In a regrettable twist of fate for human beings, many bacteria play host to specialized transposons, termed integrons, that are like flypaper in their propensity for capturing new genes. These integrons can consist of several different resistance genes, which are passed to other bacteria as whole regiments of antibiotic-defying guerrillas.

Many bacteria possessed resistance genes even before commercial antibiotics came into use. Scientists do not know exactly why these genes evolved and were maintained. A logical argument holds that natural antibiotics were initially elaborated as the result of chance genetic mutations. Then the compounds, which turned out to eliminate competitors, enabled the manufacturers to survive and proliferate—if they were also lucky enough to possess genes that protected them from their own chemical weapons. Later, these protective genes found their way into other species, some of which were pathogenic.

Regardless of how bacteria acquire resistance genes today, commercial antibiotics can select for—promote the survival and propagation of—antibiotic-resistant strains. In other words, by encouraging the growth of resistant pathogens, an antibiotic can actually contribute to its own undoing.

The selection process is fairly straightforward. When an antibiotic attacks a group of bacteria, cells that are highly susceptible to the medicine will die. But cells that have some resistance from the start, or that acquire it later (through mutation or gene exchange), may survive, especially if too little drug is given to overwhelm the cells that are present. Those cells, facing reduced competition from susceptible bacteria, will then go on to proliferate. When confronted with an antibiotic, the most resistant cells in a group will inevitably outcompete all others.

Promoting resistance in known pathogens is not the only self-defeating activity of antibiotics. When the medicines attack disease-causing bacteria, they also affect benign bacteria—innocent bystanders—in their path. They eliminate drug-susceptible bystanders that could otherwise limit the expansion of pathogens, and they simultaneously encourage the growth of resistant bystanders. Propagation of these resistant, nonpathogenic bacteria increases the reservoir of resistance traits in the bacterial population as a whole and raises the odds that such traits will spread to pathogens. In addition, sometimes the growing populations of bystanders themselves become agents of disease.

Widespread use of cephalosporin antibiotics, for example, has promoted the proliferation of the once benign intestinal bacterium *E. faecalis*, which is naturally resistant to those drugs. In most people, the immune system is able to check the growth of even multidrug-resistant *E. faecalis*, so that it does not produce illness. But in hospitalized patients with compromised immunity, the enterococcus can spread to the heart valves and other organs and establish deadly systemic disease.

Moreover, administration of vancomycin over the years has turned *E. faecalis* into a dangerous reservoir of vancomycin-resistance traits. Recall that some strains of the pathogen *S. aureus* are multidrug-resistant and are responsive only to vancomycin. Because vancomycin-resistant *E. faecalis* has become quite common, public health experts fear that it will soon deliver strong vancomycin resistance to those *S. aureus* strains, making them incurable.

The bystander effect has also enabled multidrug-resistant strains of Acinetobacter and Xanthomonas to emerge and become agents of potentially fatal blood-borne infections in hospitalized patients. These formerly innocuous microbes were virtually unheard of just a decade ago.

As I noted earlier, antibiotics affect the mix of resistant and nonresistant bacteria both in the individual being treated and in the environment. When resistant bacteria arise in treated individuals, these microbes, like other bacteria, spread readily to the surrounds and to new hosts. Investigators have shown that when one member of a household chronically takes an antibiotic to treat acne, the concentration of antibiotic-resistant bacteria on the skin of family members rises. Similarly, heavy use of antibiotics in such settings as hospitals, day care centers and farms (where the drugs are often given to livestock for nonmedicinal purposes) increases the levels of resistant bacteria in people and other organisms who are not being treated—including in individuals who live near those epicenters of high consumption or who pass through the centers.

Given that antibiotics and other antimicrobials, such as fungicides, affect the kinds of bacteria in the environment and people around the individual being treated, I often refer to these substances as societal drugs—the only class of therapeutics that can be so designated. Anticancer drugs, in contrast, affect only the person taking the medicines.

On a larger scale, antibiotic resistance that emerges in one place can often spread far and wide. The ever-increasing volume of international travel has hastened transfer to the U.S. of multidrug-resistant tuberculosis from other countries. And investigators have documented the migration of a strain of multidrug-resistant *Streptococcus pneumoniae* from Spain to the U.K., the U.S., South Africa and elsewhere. This bacterium, also known as the pneumococcus, is a cause of pneumonia and meningitis, among other diseases.

For those who understand that antibiotic delivery selects for resistance, it is not surprising that the international community currently faces a major public health crisis. Antibiotic use (and misuse) has soared since the first commercial versions were introduced and now includes many nonmedicinal applications. In 1954 two million pounds were produced in the U.S.; in 1998 the figure exceeds 50 million pounds.

Human treatment accounts for roughly half the antibiotics consumed every year in the U.S. Perhaps only half that use is appropriate, meant to cure bacterial infections and administered correctly—in ways that do not strongly encourage resistance.

Notably, many physicians acquiesce to misguided patients who demand antibiotics to treat colds and other viral infections that cannot be cured by the drugs. Researchers at the Centers for Disease Control and Prevention have estimated that some 50 million of the 150 million outpatient prescriptions for antibiotics every year are unneeded. At a seminar I conducted, more than 80 percent of the physicians present admitted to having written antibiotic prescriptions on demand against their better judgment.

In the industrial world, most antibiotics are available only by prescription, but this restriction does not ensure proper use. People often fail to finish the full course of treatment. Patients then stockpile the leftover doses and medicate themselves, or their family and friends, in less than therapeutic amounts. In both circumstances, the improper dosing will fail to eliminate the disease agent completely and will, furthermore, encourage growth of the most resistant strains, which may later produce hard-to-treat disorders. In the developing world, antibiotic use is even less controlled. Many of the same drugs marketed in the industrial nations are available over the counter. Unfortunately, when resistance becomes a clinical problem, those countries, which often do not have access to expensive drugs, may have no substitutes available.

The same drugs prescribed for human therapy are widely exploited in animal husbandry and agriculture. More than 40 percent of the antibiotics manufactured in the U.S. are given to animals. Some of that amount goes to treating or preventing infection, but the lion's share is mixed into feed to promote growth. In this last application, amounts too small to combat infection are delivered for weeks or months at a time. No one is entirely sure how the drugs support growth. Clearly, though, this long-term exposure to low doses is the perfect formula for selecting increasing numbers of resistant bacteria in the treated animals—which may then pass the microbes to care-takers and, more broadly, to people who prepare and consume undercooked meat.

In agriculture, antibiotics are applied as aerosols to acres of fruit trees, for controlling or preventing bacterial infections. High concentrations may kill all the bacteria on the trees at the time of spraying, but lingering antibiotic residues can encourage the growth of resistant bacteria that later colonize the fruit during processing and shipping. The aerosols also hit more than the targeted trees. They can be carried considerable distances to other trees and food plants, where they are too dilute to eliminate full-blown infections but are still capable of killing off sensitive bacteria and thus giving the edge to resistant versions. Here, again, resistant bacteria can make their way into people through the food chain, finding a home in the intestinal tract after the produce is eaten.

The amount of resistant bacteria people acquire from food apparently is not trivial. French researcher Denis E. Corpet showed that when human volunteers went on a diet consisting only of bacteria-free foods, the number of resistant bacteria in their feces decreased 1,000-fold. This finding suggests that we deliver a supply of resistant strains to our intestinal tract whenever we eat raw or undercooked items. These bacteria usually are not harmful, but they could be if by chance a disease-causing type contaminated the food.

The extensive worldwide exploitation of antibiotics in medicine, animal care and agriculture constantly selects for strains of bacteria that are resistant to the drugs. Must all antibiotic use be halted to stem the rise of intractable bacteria? Certainly not. But if the drugs are to retain their power over pathogens, they have to be used more responsibly. Society can accept some increase in the fraction of resistant bacteria when a disease needs to be treated; the rise is unacceptable when antibiotic use is not essential.

A number of corrective measures can be taken right now. As a start, farmers should be helped to find inexpensive alternatives for encouraging animal growth and protecting fruit trees. Improved hygiene, for instance, could go a long way to enhancing livestock development.

The public can wash raw fruit and vegetables thoroughly to clear off both resistant bacteria and possible antibiotic residues. When they receive prescriptions for antibiotics, they should complete the full course of therapy (to ensure that all the pathogenic bacteria die) and should not "save" any pills for later use. Consumers also should refrain from demanding antibiotics for colds and other viral infections and might consider seeking nonantibiotic therapies for minor conditions, such as certain cases of acne. They can continue to put antibiotic ointments on small cuts, but they should think twice about routinely using hand lotions and a proliferation of other products now imbued with antibacterial agents. New laboratory findings indicate that certain of the bacteria-fighting chemicals being incorporated into consumer products can select for bacteria resistant both to the antibacterial preparations and to antibiotic drugs.

Physicians, for their part, can take some immediate steps to minimize any resistance ensuing from required uses of antibiotics. When possible, they should try to identify the causative pathogen before beginning therapy, so they can prescribe an antibiotic targeted specifically to that microbe instead of having to choose a broad-spectrum product. Washing hands after seeing each patient is a major and obvious, but too often overlooked, precaution.

To avoid spreading multidrug-resistant infections between hospitalized patients, hospitals place the affected patients in separate rooms, where they are seen by gloved and gowned health workers and visitors. This practice should continue.

Having new antibiotics could provide more options for treatment. In the 1980s pharmaceutical manufacturers, thinking infectious diseases were essentially conquered, cut back severely on searching for additional antibiotics. At the time, if one drug failed, another in the arsenal would usually work (at least in the industrial nations, where supplies are plentiful). Now that this happy state of affairs is coming to an end, researchers are searching for novel antibiotics again. Regrettably, though, few drugs are likely to pass soon all technical and regulatory hurdles needed to reach the market. Furthermore, those that are close to being ready are structurally similar to existing antibiotics; they could easily encounter bacteria that already have defenses against them.

With such concerns in mind, scientists are also working on strategies that will give new life to existing antibiotics (see "New Technique Resensitizes Resistant Bacteria to 'Last Resort' Antibiotic," page 45). Many bacteria evade penicillin and its relatives by switching on an enzyme, penicillinase, that degrades those compounds. An antidote already on pharmacy shelves contains an inhibitor of penicillinase; it prevents the breakdown of penicillin and so frees the antibiotic to work normally. In one of the strategies under study, my laboratory is developing a compound to jam a microbial pump that ejects tetracycline from bacteria; with the pump inactivated, tetracycline can penetrate bacterial cells effectively.

As exciting as the pharmaceutical research is, overall reversal of the bacterial resistance problem will require public health officials, physicians, farmers and others to think about the effects of antibiotics in new ways. Each time an antibiotic is delivered, the fraction of resistant bacteria in the treated individual and, potentially, in others, increases. These resistant strains endure for some time—often for weeks—after the drug is removed.

The main way resistant strains disappear is by squaring off with susceptible versions that persist in—or enter—a treated person after antibiotic use has stopped. In the absence of antibiotics, susceptible strains have a slight survival advantage, because the resistant bacteria have to divert some of their valuable energy from reproduction to maintaining antibiotic-fighting traits. Ultimately, the susceptible microbes will win out, if they are available in the first place and are not hit by more of the drug before they can prevail.

Correcting a resistance problem, then, requires both improved management of antibiotic use and restoration of the environmental bacteria susceptible to these drugs. If all reservoirs of susceptible bacteria were eliminated, resistant forms would face no competition for survival and would persist indefinitely.

In the ideal world, public health officials would know the extent of antibiotic resistance in both the infectious and benign bacteria in a community. To treat a specific pathogen, physicians would favor an antibiotic most likely to encounter little resistance from any bacteria in the community. And they would deliver enough antibiotic to clear the infection completely but would not prolong therapy so much as to destroy all susceptible bystanders in the body.

Prescribers would also take into account the number of other individuals in the setting who are being treated with the same antibiotic. If many patients in a hospital ward were being given a particular antibiotic, this high density of use would strongly select for bacterial strains unsubmissive to that drug and would eliminate susceptible strains. The ecological effect on the ward would be broader than if the total amount of the antibiotic were divided among just a few people. If physicians considered the effects beyond their individual patients, they might decide to prescribe different antibiotics for different patients, or in different wards, thereby minimizing the selective force for resistance to a single medication.

Put another way, prescribers and public health officials might envision an "antibiotic threshold": a level of antibiotic usage able to correct the infections within a hospital or community but still falling below a threshold level that would strongly encourage propagation of resistant strains or would eliminate large numbers of competing, susceptible microbes. Keeping treatment levels below the threshold would ensure that the original microbial flora in a person or a community could be restored rapidly by susceptible bacteria in the vicinity after treatment ceased.

The problem, of course, is that no one yet knows how to determine where that threshold lies, and most hospitals and communities lack detailed data on the nature of their microbial populations. Yet with some dedicated work, researchers should be able to obtain both kinds of information.

Control of antibiotic resistance on a wider, international scale will require cooperation among countries around the globe and concerted efforts to educate the world's populations about drug resistance and the impact of improper antibiotic use. As a step in this direction, various groups are now attempting to track the emergence of resistant bacterial strains. For example, an international organization, the Alliance for the Prudent Use of Antibiotics has been monitoring the worldwide emergence of such strains since 1981. The group shares information with members in more than 90 countries. It also produces educational brochures for the public and for health professionals.

The time has come for global society to accept bacteria as normal, generally beneficial components of the world and not try to eliminate them—except when they give rise to disease. Reversal of resistance requires a new awareness of the broad consequences of antibiotic use—a perspective that concerns itself not only with curing bacterial disease at the moment but also with preserving microbial communities in the long run, so that bacteria susceptible to antibiotics will always be there to outcompete resistant strains. Similar enlightenment should influence the use of drugs to combat parasites, fungi and viruses. Now that consumption of those medicines has begun to rise dramatically, troubling resistance to these other microorganisms has begun to climb as well.

Fighting Bacterial Infections with Bacteria

Kristin Leutwyler

In the January 27, 2001, issue of the *British Medical Journal*, researchers describe a new way to fight bacteria with bacteria—and protect children who suffer from recurrent ear infections (otitis media). Kristian Roos and colleagues at Lundby Hospital in Gothenburg, Sweden, wondered whether infection-prone children might be treated simply by beefing up their own bodies' first line of defense: the natural flora in the upper respiratory tract. A robust colony of alpha-streptococcal bacteria, they reasoned, might run interference against pathogenic, or disease-causing, bacteria, preventing them from taking hold.

To test the idea, the researchers identified a group of 108 children with frequent ear infections between six months and six years old, and gave them all a 10-day course of antibiotics. Next they divided the children into two groups and gave them either a placebo solution or alpha-streptococcal bacteria sprayed into the nose for the next 10 days. After 60 days the kids received a second dose of either bacteria or placebo. And at the three-month mark, the scientists found that 42 percent of the children given the streptococcal spray were healthy, whereas only 22 percent of those given placebo were infection-free.

"Most antibiotics used to treat infections in the upper respiratory tract have an impact on the normal bacterial flora," the authors write. "As these bacteria are part of the body's natural defense, treatment with antibiotics abates this part of the defense system and thus facilitates colonization with pathogenic bacteria. Paradoxically, repeated courses of antibiotics might contribute to recurrent infections in children who are prone to otitis."

The Cutting Edge

(From "Behind Enemy Lines")

K.C. Nicolaou and Christopher N.C. Boddy

Bacteria find myriad strategies to outwit drugs. Obviously, the need for new, improved or even revived antibiotics is enormous. Historically, the drug discovery process identified candidates using whole-cell screening, in which molecules of interest were applied to living bacterial cells. This approach has been very successful and underlies the discovery of many drugs, including vancomycin. Its advantage lies in its simplicity and in the fact that every possible drug target in the cell is screened. But screening such a large number of targets also has a drawback. Various targets are shared by both bacteria and humans; compounds that act against those are toxic to people. Furthermore, researchers gain no information about the mechanism of action: chemists know that an agent worked, but they have no information about how. Without this critical information it is virtually impossible to bring a new drug all the way to the clinic.

Molecular-level assays provide a powerful alternative. This form of screen identifies only those compounds that have a specified mechanism of action. For instance, one such screen would look specifically for inhibitors of the transpeptidase enzyme. Although these assays are difficult to design, they yield potential drugs with known modes of action. The trouble is that only one enzyme is usually investigated at a time. It would be a vast improvement in the drug discovery process if researchers could review more than one target simultaneously, as they do in the whole-cell process, but also retain the implicit knowledge of the way the drug works. Scientists have accomplished this feat by figuring out how to assemble the many-enzyme pathway of a certain bacterium in a test tube. Using this system, they can identify molecules that either strongly disrupt one of the enzymes or subtly disrupt many of them.

Automation and miniaturization have also significantly improved the rate at which compounds can be screened. Robotics in so-called high-throughput machines allow scientists to review thousands of compounds per week. At the same time, miniaturization has cut the cost of the process by using ever-smaller amounts of reagents. In the new ultrahigh-throughput screening systems, hundreds of thousands of compounds can be looked at cost-effectively in a single day. Accordingly, chemists have to work hard to keep up with the demand for molecules. Their work is made possible by new methods in combinatorial chemistry, which allows them to design huge libraries of compounds quickly. In the future, some of these new molecules will most likely come from bacteria themselves. By understanding the way these organisms produce antibiotics, scientists can genetically engineer them to produce new related molecules.

New Technique Resensitizes Resistant Bacteria to "Last Resort" Antibiotic

Kate Wong

Over the past decade, strains of the bacteria *Staphylococcus* have grown increasingly resistant to almost all antibiotics. Most still yield to the "last resort" drug, vancomycin. But cases of related vancomycin-resistant bacteria are on the rise, leading to a growing concern that the ubiquitous *S. aureus* will develop such resistance and wreak havoc in hospitals and elsewhere. To that end, research appearing in the current issue of the journal *Science* may offer new hope in the battle against these emerging superbugs. According to the report, researchers have developed a technique that resensitizes vancomycin-resistant bacteria to the antibiotic.

Vancomycin operates by binding to the bacterial cell wall and interfering with cell wall growth. Resistant bacteria, however, have an alteration in the chemical composition of the cell wall that prevents vancomycin from binding to it. To get around that problem, Gabriela Chiosis of Columbia University and Ivo G. Boneca of Rockefeller University designed a small molecule, dubbed SProC5, that cleaves the chemically-altered cell wall component, thereby restoring the bacteria's vulnerability to vancomycin. Exposing resistant *Enterococcus faecium* (a relative of the staph bacteria) to vancomycin in combination with SProC5 effectively combatted the bacteria in mice, the team found.

Whether the approach will work in humans remains to be seen. But so far, restoring vancomycin sensitivity with molecules that cleave the resistant bacteria's altered cell wall component, the authors conclude, "is a promising strategy."

*If they are alive, they "live" in a way that defies human under-
standing of the word. Viruses are packages of genetic material
cased in a protein shell that target then hijack living cells. The
introduced genetic material reprograms the cell to produce a new
brood of viruses that escape and prowl around, looking for new
cells to infect.*

*Although they do not technically qualify as life forms, viruses
do display some very lifelike behavior. Many infectious viruses
mutate quickly, becoming resistant to what limited treatments are
now available. They freely swap genes with one another, even
jumping the barriers between animal species to create new and
unique viruses never seen before.*

*Recently, advances in genomics research and technology, some
of which are described in the following articles, have made it possible to examine in detail the entire genetic structure of a virus.
This research opens up new avenues for treatment by inhibiting a
virus's growth or replication. It also facilitates the growth of cus-
tom-built viruses where some kind of desirable genetic trait can
be grafted to a virus, injected into the body, and then allowed to
go about its single-minded task of reprogramming cells.*

Beyond Chicken Soup

William A. Haseltine

Back in the mid-1980s, when scientists first learned that a virus caused a relentless new disease named AIDS, pharmacy shelves were loaded with drugs able to treat bacterial infections. For viral diseases, though, medicine had little to offer beyond chicken soup and a cluster of vaccines. The story is dramatically different today. Dozens of antiviral therapies, including several new vaccines, are available, and hundreds more are in development. If the 1950s were the golden age of antibiotics, we are now in the early years of the golden age of antivirals.

This richness springs from various sources. Pharmaceutical companies would certainly point to the advent in the past 15 years of sophisticated techniques for discovering all manner of drugs. At the same time, frantic efforts to find lifesaving therapies for HIV, the cause of AIDS, have suggested creative ways to fight not only HIV but other viruses, too.

A little-recognized but more important force has also been at work: viral genomics, which deciphers the sequence of "letters," or nucleic acids, in a virus's genetic "text." This sequence includes the letters in all the virus's genes, which form the blueprints for viral proteins; these proteins, in turn, serve as the structural elements and the working parts of the virus and thus control its behavior. With a full or even a partial genome sequence in hand, scientists can quickly learn many details of how a virus causes disease—and which stages of the process might be particularly vulnerable to attack. In 2001 the full genome of any virus can be sequenced within days, making it possible to spot that virus's weaknesses with unprecedented speed.

The majority of antivirals on sale these days take aim at HIV, herpesviruses (responsible for a range of ills, from cold sores to encephalitis), and hepatitis B and C viruses (both of which can cause liver cancer). HIV and these forms of hepatitis will surely remain a main focus of investigation for some time; together they cause more than 250,000 cases of disease in the U.S. every year and millions in other countries. Biologists, however, are working aggressively to combat other viral illnesses as well. I cannot begin to describe all the classes of antivirals on the market and under study, but I do hope this article will offer a sense of the extraordinary advances that genomics and other sophisticated technologies have made possible in recent years.

The earliest antivirals (mainly against herpes) were introduced in the 1960s and emerged from traditional drug-discovery methods. Viruses are structurally simple, essentially consisting of genes and perhaps some enzymes (biological catalysts) encased in a protein capsule and sometimes also in a lipid envelope. Because this design requires viruses to replicate inside cells, investigators infected cells, grew them in culture and exposed the cultures to chemicals that might plausibly inhibit viral activities known at the time. Chemicals that reduced the amount of virus in the culture were considered for in-depth investigation. Beyond being a rather hit-or-miss process, such screening left scientists with few clues to other viral activities worth attacking. This handicap hampered efforts to develop drugs that were more effective or had fewer side effects.

Genomics has been a springboard for discovering fresh targets for attack and has thus opened the way to development of whole new classes of antiviral drugs. Most viral targets selected since the 1980s have been identified with the help of genomics, even though the term itself was only coined in the late 1980s, well after some of the currently available antiviral drugs were developed.

After investigators decipher the sequence of code letters in a given virus, they can enlist computers to compare that sequence with those already identified in other organisms, including other viruses, and thereby learn how the sequence is segmented into genes. Strings of code letters that closely resemble known genes in other organisms are likely to constitute genes in the virus as well and to give rise to proteins that have similar structures. Having located a virus's genes, scientists can study the functions of the corresponding proteins and thus build a comprehensive picture of the molecular steps by which the virus of interest gains a foothold and thrives in the body.

That picture, in turn, can highlight the proteins—and the domains within those proteins—that would be good to disable. In general, investigators favor targets whose disruption would impair viral activity most. They also like to focus on protein domains that bear little resemblance to those in humans, to avoid harming healthy cells and causing intolerable side effects. They take aim, too, at protein domains that are basically identical in all major strains of the virus, so that the drug will be useful against the broadest possible range of viral variants.

After researchers identify a viral target, they can enlist various techniques to find drugs that are able to perturb it. Drug sleuths can, for example, take advantage of standard genetic engineering (introduced in the 1970s) to produce pure copies of a selected protein for use in drug development. They insert the corresponding gene into bacteria or other types of cells, which synthesize endless copies of the encoded protein. The resulting protein molecules can then form the basis of rapid screening tests: only substances that bind to them are pursued further.

Alternatively, investigators might analyze the three-dimensional structure of a protein domain and then design drugs that bind tightly to that region. For instance, they might construct a compound that inhibits the active site of an enzyme crucial to viral reproduction. Drugmakers can also combine old-fashioned screening methods with the newer methods based on structures.

Advanced approaches to drug discovery have generated ideas for thwarting viruses at all stages of their life cycles. Viral species vary in the fine details of their reproductive strategies. In general, though, the stages of viral replication include attachment to the cells of a host, release of viral genes into the cells' interiors, replication of all viral genes and proteins (with help from the cells' own protein-making machinery), joining of the components into hordes of viral particles, and escape of those particles to begin the cycle again in other cells.

The ideal time to ambush a virus is in the earliest stage of an infection, before it has had time to spread throughout the body and cause symptoms. Vaccines prove their worth at that point, because they prime a person's immune system to specifically destroy a chosen disease-causing agent, or pathogen, almost as soon as it enters the body. Historically vaccines have achieved this priming by exposing a person to a killed or weakened version of the infectious agent that cannot make enough copies of itself to cause disease. So-called subunit vaccines are the most common alternative to these. They contain mere fragments of a pathogen; fragments alone have no way to produce an infection but, if selected carefully, can evoke a protective immune response.

An early subunit vaccine, for hepatitis B, was made by isolating the virus from the plasma (the fluid component of blood) of people who were infected and then purifying the desired proteins. Today a subunit hepatitis B vaccine is made by genetic engineering. Scientists use the gene for a specific hepatitis B protein to manufacture pure copies of the protein. Additional vaccines developed with the help of genomics are in development for other important viral diseases, among them dengue fever, genital herpes and the often fatal hemorrhagic fever caused by the Ebola virus.

Several vaccines are being investigated for preventing or treating HIV. But HIV's genes mutate rapidly, giving rise to many viral strains; hence, a vaccine that induces a reaction against certain strains might have no effect against others. By comparing the genomes of the various HIV strains, researchers can find sequences that are present in most of them and then use those sequences to produce purified viral protein fragments. These can be tested for their ability to induce immune protection against strains found worldwide. Or vaccines might be tailored to the HIV variants prominent in particular regions.

Treatments become important when a vaccine is not available or not effective. Antiviral treatments effect cures for some patients, but so far most of them tend to reduce the severity or duration of a viral infection. One group of therapies limits viral activity by interfering with entry into a favored cell type.

The term "entry" actually covers a few steps, beginning with the binding of the virus to some docking site, or receptor, on a host cell and ending with "uncoating" inside the cell; during uncoating, the protein capsule (capsid) breaks up, releasing the virus's genes. Entry for enveloped viruses requires an extra step. Before uncoating can occur, these microorganisms must fuse their envelope with the cell membrane or with the membrane of a vesicle that draws the virus into the cell's interior.

Several entry-inhibiting drugs in development attempt to block HIV from penetrating cells. Close examination of the

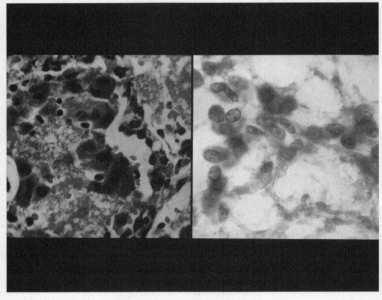

Microscopic view of AIDS related pneumonia

way HIV interacts with its favorite hosts (white blood cells called helper T cells) has indicated that it docks with molecules on those cells called CD4 and CCR5. Although blocking CD4 has failed to prevent HIV from entering cells, blocking CCR5 may yet do so.

Amantidine and rimantidine, the first two (of four) influenza drugs to be introduced, interrupt other parts of the entry process. Drugmakers found the compounds by screening likely chemicals for their overall ability to interfere with viral replication, but they have since learned more specifically that the compounds probably act by inhibiting fusion and uncoating. Fusion inhibitors discovered with the aid of genomic information are also being pursued against respiratory syncytial virus (a cause of lung disease in infants born prematurely), hepatitis B and C, and HIV.

Many colds could soon be controlled by another entry

blocker, pleconaril. Genomic and structural comparisons have shown that a pocket on the surface of rhinoviruses (responsible for most colds) is similar in most variants. Pleconaril binds to this pocket in a way that inhibits the uncoating of the virus. The drug also appears to be active against enteroviruses, which can cause diarrhea, meningitis, conjunctivitis and encephalitis.

A number of antivirals on sale and under study operate after uncoating, when the viral genome, which can take the form of DNA or RNA, is freed for copying and directing the production of viral proteins. Several of the agents that inhibit genome replication are nucleoside or nucleotide analogues, which resemble the building blocks of genes. The enzymes that copy viral DNA or RNA incorporate these mimics into the nascent strands. Then the mimics prevent the enzyme from adding any further building blocks, effectively aborting viral replication.

Acyclovir, the earliest antiviral proved to be both effective and relatively nontoxic, is a nucleoside analogue that was discovered by screening selected compounds for their ability to interfere with the replication of herpes simplex virus. It is prescribed mainly for genital herpes, but chemical relatives have value against other herpesvirus infections, such as shingles caused by varicella zoster and inflammation of the retina caused by cytomegalovirus.

The first drug approved for use against HIV, zidovudine (AZT), is a nucleoside analogue as well. Initially developed as an anticancer drug, it was shown to interfere with the activity of reverse transcriptase, an enzyme that HIV uses to copy its RNA genome into DNA. If this copying step is successful, other HIV enzymes splice the DNA into the chromosomes of an invaded cell, where the integrated DNA directs viral reproduction.

AZT can cause severe side effects, such as anemia. But studies of reverse transcriptase, informed by knowledge of the enzyme's gene sequence, have enabled drug developers to introduce less toxic nucleoside analogues. One of these, lamivudine, has also been approved for hepatitis B, which uses reverse transcriptase to convert RNA copies of its DNA genome back into DNA. Intense analyses of HIV reverse transcriptase have led as well to improved versions of a class of reverse transcriptase inhibitors that do not resemble nucleosides.

Genomics has uncovered additional targets that could be hit to interrupt replication of the HIV genome. Among these is RNase H, a part of reverse transcriptase that separates freshly minted HIV DNA from RNA. Another is the active site of integrase, an enzyme that splices DNA into the chromosomal DNA of the infected cell. An integrase inhibitor is now being tested in HIV-infected volunteers.

All viruses must at some point in their life cycle transcribe genes into mobile strands of messenger RNA, which the host cell then "translates," or uses as a guide for making the encoded proteins. Several drugs in development interfere with the transcription stage by preventing proteins known as transcription factors from attaching to viral DNA and switching on the production of messenger RNA.

Genomics helped to identify the targets for many of these agents. It also made possible a novel kind of drug: the antisense molecule. If genomic research shows that a particular protein is needed by a virus, workers can halt the protein's production by masking part of the corresponding RNA template with a custom-designed DNA fragment able to bind firmly to the selected RNA sequence. An antisense drug, fomivirsen, is already used to treat eye infections caused by cytomegalovirus in AIDS patients. And antisense agents are in development for other viral diseases; one of them blocks production of the HIV protein Tat, which is needed for the transcription of other HIV genes.

Drugmakers have also used their knowledge of viral genomes to identify sites in viral RNA that are susceptible to cutting by ribozymes—enzymatic forms of RNA. A ribozyme is being tested in patients with hepatitis C, and ribozymes for

HIV are in earlier stages of development. Some such projects employ gene therapy: specially designed genes are introduced into cells, which then produce the needed ribozymes. Other types of HIV gene therapy under study give rise to specialized antibodies that seek targets inside infected cells or to other proteins that latch onto certain viral gene sequences within those cells.

Some viruses produce a protein chain in a cell that must be spliced to yield functional proteins. HIV is among them, and an enzyme known as a protease performs this cutting. When analyses of the HIV genome pinpointed this activity, scientists began to consider the protease a drug target. With enormous help from computer-assisted structure-based research, potent protease inhibitors became available in the 1990s, and more are in development. The inhibitors that are available so far can cause disturbing side effects, such as the accumulation of fat in unusual places, but they nonetheless prolong overall health and life in many people when taken in combination with other HIV antivirals. A new generation of protease inhibitors is in the research pipeline.

Even if viral genomes and proteins are reproduced in a cell, they will be harmless unless they form new viral particles able to escape from the cell and migrate to other cells. The most recent influenza drugs, zanamivir and oseltamivir, act at this stage. A molecule called neuraminidase, which is found on the surface of both major types of influenza (A and B), has long been known to play a role in helping viral particles escape from the cells that produced them. Genomic comparisons revealed that the active site of neuraminidase is similar among various influenza strains, and structural studies enabled researchers to design compounds able to plug that site. The other flu drugs act only against type A.

Developers are also selecting novel drugs based on their ability to combat viral strains that are resistant to other drugs. Drugs can prevent the cell-to-cell spread of viruses in a different way—by augmenting a patient's immune responses. Some of these responses are nonspecific: the drugs may restrain the spread through the body of various kinds of invaders rather than homing in on a particular pathogen. Molecules called interferons take part in this type of immunity, inhibiting protein synthesis and other aspects of viral replication in infected cells. For that reason, one form of human interferon, interferon alpha, has been a mainstay of therapy for hepatitis B and C. (For hepatitis C, it is used with an older drug, ribavirin.) Other interferons are under study, too.

More specific immune responses include the production of standard antibodies, which recognize some fragment of a protein on the surface of a viral invader, bind to that protein and mark the virus for destruction by other parts of the immune system. Once researchers have the gene sequence encoding a viral surface protein, they can generate pure, or "monoclonal," antibodies to selected regions of the protein. One monoclonal is on the market for preventing respiratory syncytial virus in babies at risk for this infection; another is being tested in patients suffering from hepatitis B.

Comparisons of viral and human genomes have suggested yet another antiviral strategy. A number of viruses, it turns out, produce proteins that resemble molecules involved in the immune response. Moreover, certain of those viral mimics disrupt the immune onslaught and thus help the virus to evade destruction. Drugs able to intercept such evasion-enabling proteins may preserve full immune responses and speed the organism's recovery from numerous viral diseases. The hunt for such agents is under way.

The pace of antiviral drug discovery is nothing short of breathtaking, but at the same time, drugmakers have to confront a hard reality: viruses are very likely to develop resistance, or insensitivity, to many drugs. Resistance is especially probable when the compounds are used for long periods, as they are in such chronic diseases as HIV and in quite a few cases of hepatitis B and C. Indeed, for every HIV drug in the present arsenal, some viral strain exists that is resistant to it and, often, to additional drugs. This resistance stems from the tendency of viruses—especially RNA viruses and most especially HIV—to mutate rapidly. When a mutation enables a viral strain to overcome some obstacle to reproduction (such as a drug), that strain will thrive in the face of the obstacle. To keep the resistance demon at bay until effective vaccines are found, pharmaceutical companies will have to develop more drugs. When mutants resistant to a particular drug arise, reading their genetic text can indicate where the mutation lies in the viral genome and suggest how that mutation might alter the interaction between the affected viral protein and the drug. Armed with that information, researchers can begin structure-based or other studies designed to keep the drug working despite the mutation.

Pharmaceutical developers are also selecting novel drugs based on their ability to combat viral strains that are resistant to other drugs. Recently, for instance, DuPont Pharmaceuticals chose a new HIV nonnucleoside reverse transcriptase inhibitor, DPC 083, for development precisely because of its ability to overcome viral resistance to such inhibitors. The company's researchers first examined the mutations in the reverse transcriptase gene that conferred resistance. Next they turned to computer modeling to find drug designs likely to inhibit the reverse transcriptase enzyme in spite of those mutations. Then, using genetic engineering, they created viruses that produced the mutant enzymes and selected the compound best able to limit reproduction by those viruses. The drug is now being evaluated in HIV-infected patients.

It may be some time before virtually all serious viral infections are either preventable by vaccines or treatable by some effective drug therapy. But now that the sequence of the human genome is available in draft form, drug designers will identify a number of previously undiscovered proteins that stimulate the production of antiviral antibodies or that energize other parts of the immune system against viruses. The insights gleaned from the human genome, viral genomes and other advanced drug-discovery methods are sure to provide a flood of needed antivirals within the next 10 to 20 years.

ling bacteria is like hitt:
rget--just when you think y
em, they mutate on you. Thro
ruse and misuse of antibiot
d far___ ___ _____tingly s
elerated the development o
teria that survive convent:
c atta___ ___ ____ drugs po
e medical community preach
antibiotics and teach pati

Muscular Again

Glenn Zorpette

With relatively few old-timers showing an inclination to pump iron three times a week for the rest of their lives, the potential market for an alternative muscle-building drug is clearly enormous. And science finally appears close to creating one. In separate experiments at the University of Pennsylvania Medical Center in Philadelphia and at the Royal Free and University College Medical School in London, researchers tested muscle-building vaccines based on engineered genes. Injected into mice, the vaccines boosted muscle mass in the animals' legs by 15 to 27 percent. Amazingly, the increases were measurable in only a month or so and didn't require any exercise at all.

One of the most important growth factors is insulinlike growth factor-1 (IGF-1). During infancy and childhood, IGF-1 produced by the liver circulates throughout the body, rapidly expanding all the body's muscle fibers. The amount of this circulating, liver-produced IGF-1 eventually declines sharply, ending the early-life growth spurt. For muscle growth, the free ride is then over, and only exercise can add (and eventually, merely maintain) muscle mass. IGF-1 and other growth factors continue to play a major role, but they are released only locally in muscle during exercise or in response to injury.

It was this local, muscle-specific form of IGF-1 that the University of Pennsylvania researchers exploited in their genetic experiments on mice. The Penn team, led by H. Lee Sweeney, took the gene that codes for the rodent form of muscle-specific IGF-1 and put it in a virus. Viruses can be useful for splicing engineered genes into cells because they target the nucleus, inserting the genes into a chromosome so that the DNA is not lost over time.

Injected into a mouse's leg, the virus eventually got into 50 to 75 percent of the leg's muscle cells, Sweeney estimates. In each cell, the virus entered the cytoplasm and broke up, releasing the engineered gene and the associated genetic material. By mechanisms not well understood, the gene and other DNA became integrated into the nucleus's own DNA. Intriguingly, the invading DNA seemed to position itself randomly on a chromosome, Sweeney reports.

Once on the chromosome, with its promoter region stuck in the on position, the engineered gene started transcribing mRNA for muscle-specific IGF-1. The transcription continued until the animal died, of old age.

Lest couch potatoes rejoice, several major obstacles would have to be overcome before injections let inactive senior citizens go from park benches to bench presses. Still, many muscle researchers believe that human tests are inevitable, and some think the first ones will occur within the next couple of years. Not only would such a vaccine be about as close as humanity is likely to come anytime soon to an anti-aging elixir, but it could also be a major breakthrough for the treatment of a host of degenerative muscle diseases, including the various forms of muscular dystrophy.

Many Ways to Make a Vaccine

(From "Better than a Cure")

Tim Beardsley

As long ago as the early 18th century, people sought protection against smallpox by pressing infectious "matter" from patients' lesions into breaks in their own skin. The highly dangerous practice was intended to cause a mild case of the disease and so bestow immunity.

Immunization became a more reasonable proposition in 1798, when Edward Jenner demonstrated how the illness could be prevented by inoculations of cowpox, a related but less dangerous disease. The principle of using a related pathogen to provoke immunity is still being explored today in vaccines designed to protect children from rotaviruses, which cause often fatal diarrhea in millions of children. But many other approaches to vaccination are used as well.

Killed whole organisms and weakened toxins

Several widely used vaccines, including those targeted at influenza and pertussis, are based on killed microbes. Injectable poliomyelitis vaccines also work this way. A variant of this approach, used in diphtheria and tetanus immunization, is to inject people with chemically modified versions of toxins produced by the infectious agent. Such toxoids, as they are called, allow the immune system to learn how to inactivate the poison from a real infection.

Subunit vaccines

Vaccines consisting of molecular subunits of pathogens can avoid some of the complications of using whole organisms.

Dr. Albert Sabin examining a vial that holds a live-virus vaccine for polio

Subunit vaccines are now available for meningitis, pneumonia and hepatitis B; vaccines employing this approach against respiratory syncytial virus and parainfluenza virus, both major killers, are in development. Subunit preparations are also being investigated as candidates for protection against infection with HIV (the AIDS virus) as well as malaria.

Altered pathogens

Some diseases require a more powerful immune stimulus. In such cases, favorable results can sometimes be obtained with live microorganisms that have been weakened, or attenuated, so that they do not produce significant illness. This is still the basis of the oral polio vaccine and the combined vaccine against measles, mumps and rubella. Modified pathogens might be even more commonly used in the future. Genetic engineering has made it possible to introduce immune-stimulating proteins from a range of pathogens into tried and trusted carrier organisms such as vaccinia (derived from the cowpox virus with which Jenner countered smallpox) and various bacteria.

Conjugates

Vaccines for some bacterial diseases, such as pneumococcal pneumonia and meningitis, cannot be used in babies, because their immune systems do not recognize as foreign the sugars in the bacterial cell walls. During the past two decades, however, researchers have learned how to combine these sugars with protein carriers. The resulting "conjugate" vaccines work well even in infants. In 1986 the first conjugate vaccine, against Hemophilus influenzae type B, was licensed in the U.S., and current versions are effective in children as young as two months. Conjugate vaccines for pneumococcal pneumonia and meningococci groups A and C are under development. "They could have a tremendous impact in a number of diseases," says John R. La Montagne of the National Institute of Allergy and Infectious Diseases.

Adjuvants and microspheres

Many improvements in vaccines expected over the next decade are likely to be the result of better adjuvants or carriers. Adjuvants are substances that potentiate an immune response. The only adjuvant licensed for use in humans at present is alum, an aluminum salt, but other chemicals, including some complex organic ones, are being studied. Among them are muramyl dipeptide, squalene, lipid spheres known as liposomes and cagelike organic structures called immunostimulatory complexes (ISCOMs).

An approach that has long captured the imagination of researchers puts immunogenic chemical fragments into minute polymer spheres that only slowly let the immune system "see" their contents as they diffuse out. The scheme might be able to confer long-lasting immunity for some diseases with a single inoculation. Recently workers have successfully stabilized tetanus toxoid in polyester microspheres, thus pointing the way to a one-dose tetanus vaccine.

Oral vaccines

Most vaccines are thought to exert their effects in the bloodstream. Oral polio vaccine, however, seems to work by triggering a different kind of immune response that can be elicited only when immunostimulatory molecules reach special cells in the lining of the gut. Immunity generated this way is termed mucosal immunity. Many researchers believe vaccines designed to stimulate mucosal immunity—which would probably be administered orally—might protect against diseases that have so far proved resistant to vaccination. Sexually transmitted diseases, including HIV infection, are a major focus of research interest, as are pathogens that enter the body through the gut, such as Vibrio cholerae (which causes cholera) and Shigella (dysentery).

Naked DNA

The most recent and surprising development was the demonstration that DNA, when injected into muscle, can by itself confer immunity. Workers at Vical, a biotechnology company in San Diego, Calif., collaborating with Merck and U.S. Navy investigators, have shown that such "naked DNA" can immunize mice against malaria and influenza. The DNA, which encodes a protein displayed by the pathogen, apparently stimulates host tissues to synthesize proteins that the immune system recognizes as foreign. By continuously stimulating the immune system, naked DNA vaccines could produce responses as strong as those induced by attenuated organisms. In particular, they seem able to stimulate the arm of the immune system that employs T cells to kill invaders. Naked DNA technology "has an immensely powerful capability," says Philip K. Russell of the Albert B. Sabin Vaccine Foundation.

Some Vaccines in Development (1995)

Disease/Pathogen	Technology
Respiratory syncytial virus	Attenuated virus; subunit in microspheres
Influenza	Attenuated virus and naked DNA
Group B streptococci	Conjugates
Parainfluenza	Subunits and attenuated virus; inactivated virus in microspheres
Meningococci group B	Modified polysaccharide
Measles	Subunits in ISCOMs and in vaccinia; attenuated virus; naked DNA

Some Vaccines in Development (continued)

Disease/Pathogen	*Technology*
Pneumococcal pneumonia	Subunits and conjugates
Typhoid	Subunits and conjugates
Cholera	Inactivated and attenuated pathogens; subunit combinations
Shigella	Bacterial vector; subunits and conjugates
Tuberculosis	Mycobacterial vector; subunit
Malaria	Subunit and naked DNA
Dengue	Yeast, yellow fever and vaccinia vectors; attenuated and chimeric viruses
Rotavirus	Attenuated and modified viruses
Tetanus	Single dose: toxoid in microspheres
Schistosomiasis	Subunit
Human Immunodeficiency Virus	Subunits; vaccinia vectors; inactivated virus

SOURCES: Global Program for Vaccines and Immunization, World Health Organization; The Jordan Report, National Institute of Allergy and Infectious Diseases

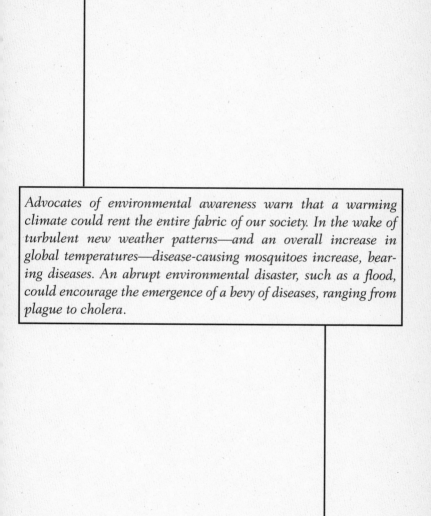

Advocates of environmental awareness warn that a warming climate could rent the entire fabric of our society. In the wake of turbulent new weather patterns—and an overall increase in global temperatures—disease-causing mosquitoes increase, bearing diseases. An abrupt environmental disaster, such as a flood, could encourage the emergence of a bevy of diseases, ranging from plague to cholera.

Is Global Warming Harmful to Health?

Paul R. Epstein

Today few scientists doubt the atmosphere is warming. Most also agree that the rate of heating is accelerating and that the consequences of this temperature change could become increasingly disruptive. Even high school students can reel off some projected outcomes: the oceans will warm, and glaciers will melt, causing sea levels to rise and salt water to inundate settlements along many low-lying coasts. Meanwhile the regions suitable for farming will shift. Weather patterns should also become more erratic and storms more severe.

Global warming can also threaten human well-being profoundly, if somewhat less directly, by revising weather patterns—particularly by pumping up the frequency and intensity of floods and droughts and by causing rapid swings in the weather. As the atmosphere has warmed over the past century, droughts in arid areas have persisted longer, and massive bursts of precipitation have become more common. Aside from causing death by drowning or starvation, these disasters promote by various means the emergence, resurgence and spread of infectious disease.

That prospect is deeply troubling, because infectious illness is a genie that can be very hard to put back into its bottle. It may kill fewer people in one fell swoop than a raging flood or an extended drought, but once it takes root in a community, it often defies eradication and can invade other areas.

The control issue looms largest in the developing world, where resources for prevention and treatment can be scarce. But the technologically advanced nations, too, can fall victim to surprise attacks—as happened last year when the West Nile virus broke out for the first time in North America, killing seven New Yorkers (see "Outbreak Not Contained," page 83). In these days of international commerce and travel, an infectious disorder that appears in one part of the world can quickly become a problem continents away if the disease-causing agent, or pathogen, finds itself in a hospitable environment.

Floods and droughts associated with global climate change could undermine health in other ways as well. They could damage crops and make them vulnerable to infection and infestations by pests and choking weeds, thereby reducing food supplies and potentially contributing to malnutrition. And they could permanently or semipermanently displace entire populations in developing countries, leading to overcrowding and the diseases connected with it, such as tuberculosis.

Diseases relayed by mosquitoes—such as malaria, dengue fever, yellow fever and several kinds of encephalitis—are among those eliciting the greatest concern as the world warms. Mosquitoes acquire disease-causing microorganisms when they take a blood meal from an infected animal or person. Then the pathogen reproduces inside the insects, which may deliver disease-causing doses to the next individuals they bite.

Mosquito-borne disorders are projected to become increasingly prevalent because their insect carriers, or "vectors," are very sensitive to meteorological conditions. Cold can be a friend to humans, because it limits mosquitoes to seasons and regions where temperatures stay above certain minimums.

Dr. George Kido supervising the use of his mosquito control machine in 1943

Winter freezing kills many eggs, larvae and adults outright. Anopheles mosquitoes, which transmit malaria parasites (such as *Plasmodium falciparum*), cause sustained outbreaks of malaria only where temperatures routinely exceed 60 degrees Fahrenheit. Similarly, Aedes aegypti mosquitoes, responsible for yellow fever and dengue fever, convey virus only where temperatures rarely fall below 50 degrees F.

Excessive heat kills insects as effectively as cold does. Nevertheless, within their survivable range of temperatures, mosquitoes proliferate faster and bite more as the air becomes warmer. At the same time, greater heat speeds the rate at which pathogens inside them reproduce and mature. At 68 degrees F, the immature *P. falciparum* parasite takes 26 days to develop fully, but at 77 degrees F, it takes only 13 days. The Anopheles mosquitoes that spread this malaria parasite live only several weeks; warmer temperatures raise the odds that the parasites will mature in time for the mosquitoes to transfer the infection. As whole areas heat up, then, mosquitoes could expand into formerly forbidden territories, bringing illness with them. Further, warmer nighttime and winter temperatures may enable them to cause more disease for longer periods in the areas they already inhabit.

The extra heat is not alone in encouraging a rise in mosquito-borne infections. Intensifying floods and droughts resulting from global warming can each help trigger outbreaks by creating breeding grounds for insects whose dessicated eggs remain viable and hatch in still water. As floods recede, they leave puddles. In times of drought, streams can become stagnant pools, and people may put out containers to catch water; these pools and pots, too, can become incubators for new mosquitoes. And the insects can gain another boost if climate change or other processes (such as alterations of habitats by humans) reduce the populations of predators that normally keep mosquitoes in check.

Malaria and dengue fever are two of the mosquito-borne diseases most likely to spread dramatically as global temperatures head upward. Malaria (marked by chills, fever, aches and anemia) already kills 3,000 people, mostly children, every day. Some models project that by the end of the 21st century, ongoing warming will have enlarged the zone of potential malaria transmission from an area containing 45 percent of the world's population to an area containing about 60 percent. That news is bad indeed, considering that no vaccine is available and that the causative parasites are becoming resistant to standard drugs (see "The Challenge of Antibiotic Resistance," page 27).

True to the models, malaria is reappearing north and south

of the tropics. The U.S. has long been home to Anopheles mosquitoes, and malaria circulated here decades ago. By the 1980s mosquito-control programs and other public health measures had restricted the disorder to California. Since 1990, however, when the hottest decade on record began, outbreaks of locally transmitted malaria have occurred during hot spells in Texas, Florida, Georgia, Michigan, New Jersey and New York (as well as in Toronto). These episodes undoubtedly started with a traveler or stowaway mosquito carrying malaria parasites. But the parasites clearly found friendly conditions in the U.S.— enough warmth and humidity, and plenty of mosquitoes able to transport them to victims who had not traveled. Malaria has returned to the Korean peninsula, parts of southern Europe and the former Soviet Union and to the coast of South Africa along the Indian Ocean.

Dengue, or "breakbone," fever (a severe flulike viral illness that sometimes causes fatal internal bleeding) is spreading as well. Today it afflicts an estimated 50 million to 100 million in the tropics and subtropics (mainly in urban areas and their surroundings). It has broadened its range in the Americas over the past 10 years and had reached down to Buenos Aires by the end of the 1990s. It has also found its way to northern Australia. Neither a vaccine nor a specific drug treatment is yet available.

Although these expansions of malaria and dengue fever certainly fit the predictions, the cause of that growth cannot be traced conclusively to global warming. Other factors could have been involved as well—for instance, disruption of the environment in ways that favor mosquito proliferation, declines in mosquito-control and other public health programs, and rises in drug and pesticide resistance. The case for a climatic contribution becomes stronger, however, when other projected consequences of global warming appear in concert with disease outbreaks.

The increased climate variability accompanying warming

will probably be more important than the rising heat itself in fueling unwelcome outbreaks of certain vector-borne illnesses. For instance, warm winters followed by hot, dry summers (a pattern that could become all too familiar as the atmosphere heats up) favor the transmission of St. Louis encephalitis and other infections that cycle among birds, urban mosquitoes and humans.

This sequence seems to have abetted the surprise emergence of the West Nile virus in New York City last year. No one knows how this virus found its way into the U.S. But one reasonable explanation for its persistence and amplification here centers on the weather's effects on Culex pipiens mosquitoes, which accounted for the bulk of the transmission. These urban dwellers typically lay their eggs in damp basements, gutters, sewers and polluted pools of water.

The interaction between the weather, the mosquitoes and the virus probably went something like this: The mild winter of 1998–99 enabled many of the mosquitoes to survive into the spring, which arrived early. Drought in spring and summer concentrated nourishing organic matter in their breeding areas and simultaneously killed off mosquito predators, such as lacewings and ladybugs, that would otherwise have helped limit mosquito populations. Drought would also have led birds to congregate more, as they shared fewer and smaller watering holes, many of which were frequented, naturally, by mosquitoes.

Once mosquitoes acquired the virus, the heat wave that accompanied the drought would speed up viral maturation inside the insects. Consequently, as infected mosquitoes sought blood meals, they could spread the virus to birds at a rapid clip. As bird after bird became infected, so did more mosquitoes, which ultimately fanned out to infect human beings. Torrential rains toward the end of August provided new puddles for the breeding of C. pipiens and other mosquitoes, unleashing an added crop of potential virus carriers.

Like mosquitoes, other disease-conveying vectors tend to be "pests"—opportunists that reproduce quickly and thrive under disturbed conditions unfavorable to species with more specialized needs. In the 1990s climate variability contributed to the appearance in humans of a new rodent-borne ailment: the hantavirus pulmonary syndrome, a highly lethal infection of the lungs. This infection can jump from animals to humans when people inhale viral particles hiding in the secretions and excretions of rodents. The sequential weather extremes that set the stage for the first human eruption, in the U.S. Southwest in 1993, were long-lasting drought interrupted by intense rains.

First, a regional drought helped to reduce the pool of animals that prey on rodents—raptors (owls, eagles, prairie falcons, red-tailed hawks and kestrels), coyotes and snakes. Then, as drought yielded to unusually heavy rains early in 1993, the rodents found a bounty of food, in the form of grasshoppers and piñon nuts. The resulting population explosion enabled a virus that had been either inactive or isolated in a small group to take hold in many rodents. When drought returned in summer, the animals sought food in human dwellings and brought the disease to people. By fall 1993, rodent numbers had fallen, and the outbreak abated.

Subsequent episodes of hantavirus pulmonary syndrome in the U.S. have been limited, in part because early-warning systems now indicate when rodent-control efforts have to be stepped up and because people have learned to be more careful about avoiding the animals' droppings. But the disease has appeared in Latin America, where some ominous evidence suggests that it may be passed from one person to another.

As the natural ending of the first hantavirus episode demonstrates, ecosystems can usually survive occasional extremes. They are even strengthened by seasonal changes in weather conditions, because the species that live in changeable climates have to evolve an ability to cope with a broad range of conditions. But long-lasting extremes and very wide fluctuations in weather can overwhelm ecosystem resilience. (Persistent ocean heating, for instance, is menacing coral reef systems, and drought-driven forest fires are threatening forest habitats.) And ecosystem upheaval is one of the most profound ways in which climate change can affect human health. Pest control is one of nature's underappreciated services to people; well-functioning ecosystems that include diverse species help to keep nuisance organisms in check. If increased warming and weather extremes result in more ecosystem disturbance, that disruption may foster the growth of opportunist populations and enhance the spread of disease.

Beyond exacerbating the vector-borne illnesses mentioned above, global warming will probably elevate the incidence of waterborne diseases, including cholera (a cause of severe diarrhea). Warming itself can contribute to the change, as can a heightened frequency and extent of droughts and floods. It may seem strange that droughts would favor waterborne disease, but they can wipe out supplies of safe drinking water and concentrate contaminants that might otherwise remain dilute. Further, the lack of clean water during a drought interferes with good hygiene and safe rehydration of those who have lost large amounts of water because of diarrhea or fever.

Floods favor waterborne ills in different ways. They wash sewage and other sources of pathogens (such as *Cryptosporidium*) into supplies of drinking water. They also flush fertilizer into water supplies. Fertilizer and sewage can each combine with warmed water to trigger expansive blooms of harmful algae. Some of these blooms are directly toxic to humans who inhale their vapors; others contaminate fish and shellfish, which, when eaten, sicken the consumers. Recent discoveries have revealed that algal blooms can threaten human health in yet another way. As they grow bigger, they support the proliferation of various pathogens, among them *Vibrio cholerae*, the causative agent of cholera.

Drenching rains brought by a warmed Indian Ocean to the

Horn of Africa in 1997 and 1998 offer an example of how people will be affected as global warming spawns added flooding.
The downpours set off epidemics of cholera as well as two
mosquito-borne infections: malaria and Rift Valley fever (a flulike disease that can be lethal to livestock and people alike).
The health toll taken by global warming will depend to a large
extent on the steps taken to prepare for the dangers. The ideal
defensive strategy would have multiple components.

One would include improved surveillance systems that
would promptly spot the emergence or resurgence of infectious diseases or the vectors that carry them. Discovery could
quickly trigger measures to control vector proliferation without
harming the environment, to advise the public about self-pro-
tection, to provide vaccines (when available) for at-risk populations and to deliver prompt treatments.

Sadly, however, comprehensive surveillance plans are not
yet realistic in much of the world. And even when vaccines or
effective treatments exist, many regions have no means of
obtaining and distributing them. Providing these preventive
measures and treatments should be a global priority.

A second component would focus on predicting when cli-
matological and other environmental conditions could become
conducive to disease outbreaks, so that the risks could be minimized. If climate models indicate that floods are likely in a
given region, officials might stock shelters with extra supplies.
Or if satellite images and sampling of coastal waters indicate
that algal blooms related to cholera outbreaks are beginning,
officials could warn people to filter contaminated water and
could advise medical facilities to arrange for additional staff,
beds and treatment supplies.

A third component of the strategy would attack global
warming itself. Human activities that contribute to the heating
or that exacerbate its effects must be limited. Little doubt
remains that burning fossil fuels for energy is playing a significant role in global warming, by spewing carbon dioxide and

other heat-absorbing, or "greenhouse," gases into the air. Cleaner energy sources must be put to use quickly and broadly, both in the energy-guzzling industrial world and in developing nations, which cannot be expected to cut back on their energy use. (Providing sanitation, housing, food, refrigeration and indoor fires for cooking takes energy, as do the pumping and purification of water and the desalination of seawater for irrigation.) In parallel, forests and wetlands need to be restored, to absorb carbon dioxide and floodwaters and to filter contaminants before they reach water supplies.

The world's leaders, if they are wise, will make it their business to find a way to pay for these solutions. Climate, ecological systems and society can all recoup after stress, but only if they are not exposed to prolonged challenge or to one disruption after another. The Intergovernmental Panel on Climate Change, established by the United Nations, calculates that halting the ongoing rise in atmospheric concentrations of greenhouse gases will require a whopping 60 to 70 percent reduction in emissions.

Climate does not necessarily change gradually. The multiple factors that are now destabilizing the global climate system could cause it to jump abruptly out of its current state. At any time, the world could suddenly become much hotter or even much colder. Such a sudden, catastrophic change is the ultimate health risk—one that must be avoided at all costs.

Although long predicted, the reality of a lethal pathogen from a distant corner of the globe making its way into a crowded North American city and quickly taking a foothold happened in 1999. In that year, New York City found itself home to a new immigrant, the West Nile Virus. It killed scores of birds and seven people before health officials knew they had an outbreak on their hands. Following is the story.

Outbreak Not Contained

Marguerite Holloway

The appearance of West Nile virus in New York City in 1999 caught the U.S. by surprise. That this virus—which is known in Africa, Asia and, increasingly, in parts of Europe—could find its way to American shores and perform its deadly work for many months before being identified has shaken up the medical community. It has revealed several major gaps in the public health infrastructure that may become ever more important in this era of globalization and emerging diseases.

Because it is mosquito-borne, West Nile has reinforced the need for mosquito surveillance—something that is only sporadically practiced around the country and something that could perhaps help doctors identify other agents causing the many mysterious cases of encephalitis that occur every year. And because it killed birds before it killed seven people, the virus made dramatically clear that the cultural divide between the animal-health and the public-health communities is a dangerous one. "It was a tremendous wake-up call for the United States in general," says William K. Reisen of the Center for

Vector-Borne Disease Research at the University of California at Davis.

No one is certain when, or how, West Nile arrived in New York. The virus—one of 10 in a family called flaviviruses, which includes St. Louis encephalitis—could have come via a bird, a mosquito that had survived an intercontinental flight or an infected traveler. It is clear, however, that West Nile started felling crows in New York's Queens County in June of 1999 and had moved into the Bronx by July, where it continued to kill crows and then, in September, birds at the Bronx Zoo.

By the middle of August, people were succumbing as well. In two weeks Deborah S. Asnis, chief of infectious disease at the Flushing Hospital Medical Center in Queens, saw eight patients suffering similar neurological complaints. After the third case, and despite some differences in their symptoms, Asnis alerted the New York City Department of Health. The health department, in turn, contacted the state and the Centers for Disease Control and Prevention (CDC), and the hunt for the pathogen was on. It was first identified as St. Louis encephalitis, which has a similar clinical profile and cross-reacts with West Nile in the lab.

Understandable as it is to many health experts, the initial misidentification remains worrisome. As Reisen points out, diagnostic labs can only look for what they know. If they don't have West Nile reagents on hand, they won't find the virus, just its relatives. "In California we have had only one flavivirus that we were looking for, so if West Nile had come in five years ago, we would have missed it until we had an isolate of the virus as well," Reisen comments.

This is true even though California, unlike New York State, has an extensive, $70-million-a-year mosquito surveillance and control system. The insects are trapped every year so that their populations can be assessed and tested for viruses. Surveillance has allowed California to document the appearance of three new species of mosquito in the past 15 years. In addition,

200 flocks of 10 sentinel chickens are stationed throughout the state. Every few weeks during the summer they are tested for viral activity.

In 1990 sentinel chickens in Florida detected St. Louis encephalitis before it infected people. "Six weeks before the human cases, we knew we had a big problem," recalls Jonathan F. Day of the Florida Medical Entomology Laboratory. After warning people to take precautions and spraying with insecticides, the state documented 226 cases and 11 deaths. "It is very difficult to say how big the problem would have been if we hadn't known," Day says. "But without our actions I think it would have been in the thousands." (Day says surveillance in his county costs about $35,000 annually.)

New York City, home to perhaps about 40 species of mosquito, has no such surveillance in place, even though some of its neighbors—Suffolk County, Nassau County and every county in New Jersey—do. And it is not alone. Many cities do not monitor for the whining pests unless they are looking for a specific disease. "We have clearly forgotten about mosquito-borne disease," says Thomas P. Monath, vice president of research and medical affairs at Ora Vax in Cambridge, Mass., and formerly of the CDC. "We have let our infrastructure decay, and we have fewer experts than we had 20 or 30 years ago."

Tracking mosquitoes could potentially help not just with exotic arrivals but with the plethora of encephalitis cases reported every year. Indeed, the particular strain of West Nile that hit New York was ultimately identified by Ian Lipkin of the University of California at Irvine and his colleagues because they were collaborating with the New York State Department of Health on an encephalitis project. Two thirds of the cases of encephalitis that occur every year have an "unknown etiology." A few states, including New York, California and Tennessee, have recently started working with the CDC to develop better tests to identify some of these mysterious origins. As a result,

Lipkin—who says he has developed an assay that can quickly identify pathogens even if they are not being looked for—was given samples from the patients who had died in New York City.

Even if surveillance can't catch what it doesn't know, it can tell public health researchers that a new mosquito species has appeared—say, one that can transmit dengue or yellow fever— or it can indicate that something is wrong with the birds and should be investigated. The sentinels in the case of West Nile were, in fact, the city's crows and, later, birds at the Bronx Zoo. Through careful analysis of the crows and other species, Tracey McNamara, a veterinary pathologist at the Wildlife Conservation Society (which runs the Bronx Zoo), quickly determined that the pathogen was not St. Louis encephalitis— despite the CDC claims—because that disease does not kill birds. And she knew that it was not eastern equine encephalitis, because emus weren't dying. "We owe a debt of gratitude to the emu flock," McNamara says.

But despite her recognition that something new, unusual and deadly was afoot, McNamara could do little herself— except hound people in the human-health community to take a look at the wildlife. "The thing that was so frustrating was that we lack the infrastructure to respond," she says. "There was no vet lab in the country that could do the testing." Because none of the veterinary or wildlife labs had the ability to deal with such pathogens, McNamara was forced to send her bird samples to the CDC and to a U.S. Army lab. The Wildlife Conservation Society recently gave $15,000 to Robert G. McLean, director of the U.S. Geological Survey's National Wildlife Health Center, so he could study the pathogenesis of West Nile virus in crows and the effectiveness of an avian vaccine. "The federal budget moves at glacial speed," McNamara complains. "That is going to need to be addressed."

The continued bird work by McLean and others has kept the East Coast on alert for the potential of another West Nile outbreak. In 1999, McLean and his colleagues found West Nile in a crow in Baltimore and in a migratory bird, the eastern phoebe. "They go to the southern U.S.," he notes. "That just convinces us that a lot of migratory birds were infected and flew south with the virus." Despite the fact that "wildlife is a good warning system for what could eventually cause problems in humans," McLean is not optimistic about a true and equal collaboration between his and McNamara's world and the CDC's: "We are on the outside looking in. We are not partners yet, and I am not sure we will ever get to be partners." The cost could be high. As McNamara points out, "Don't you want a diagnosis in birds before it gets to humans?"

In Malaysia in 1999, an unknown and terrifying virus turned its victims into mental vegetables before finally killing them by poisoning their brains. It seemed to slay anyone who harbored it. More disturbingly, the virus, later named Nipan, seems to have "jumped species," starting with pigs and somehow mutating enough to lodge itself in humans. The following article details how scientists uncovered its origins.

Trailing a Virus

W. Wayt Gibbs

Chua Kaw Bing endured the 18-hour plane ride from Kuala Lumpur to Los Angeles uneasily. He hated long flights. Since he had given up his private practice to study viral outbreaks with Lam Sai Kit, a world-renowned expert on the subject and head of the University of Malaya's department of medical microbiology, the young doctor always seemed to be flying somewhere. "There is no individualism in our fight against emerging diseases—only internationalism" was the motto Lam had tacked to his whiteboard. He put the words to work in February 1999 when he dispatched Chua to Perth to get Australian help in confirming the cause of an outbreak that had painfully swelled the joints of 27 feverish people in Port Klang. Now, hardly a month later, Chua had set out on the track of a new epidemic—this time to the Centers for Disease Control and Prevention (CDC) laboratories in Fort Collins, Colo. He was counting on their high-tech equipment to identify what he could not: a mysterious and deadly virus packed carefully inside the carry-on bag at his feet.

As Chua's plane had climbed away from Kuala Lumpur's new airport toward Taipei, passengers on the left side of the jet might have just spotted the horse and swine stables near Ipoh where this strange disease had started last September. Even before it spread, it had seemed frightening enough—with 26 victims, it was the biggest outbreak of Japanese encephalitis in Malaysia in more than 25 years, they said.

And then the virus had jumped. There, to the palm-covered state of Negri Sembilan, now off the right side of the plane, the heart of Malaysian pig country. No doubt some desperate farmer, under the cover of darkness, had found a hole in the quarantine and sent his pigs south from Ipoh for sale or for slaughter. How could a simple farmer have known the biological and economic firestorm it would ignite? How could he have foreseen hundreds of people—strong men, mostly—burning with fever, slipping into delirium, coma and beyond; entire villages emptied in a panic as one household in three is touched by the disease; gas-masked soldiers opening fire on herds of swine, decimating a huge export industry farm by farm; and other farmers like him smuggling pigs through other roadblocks into other states, a chain reaction with no clear end?

The government scientists were saying the disease was Japanese encephalitis, after all, and JE is easily stopped with a vaccine. Besides, pigs may provide a host in which the JE virus can multiply, but hogs do not transmit it directly to humans: mosquitoes do. Antimosquito fogging and mass JE vaccination had always quenched JE outbreaks before, and the government had already started this.

But if there is a lesson for the world to learn from the affliction visited on Malaysia's pig farmers this past spring, perhaps it is that a new disease can look, even to the best doctors, like a familiar one. When it does, the pathogen gains time to spread. And if the malady is transmitted by a valuable commodity such as the pig rather than by a pest such as the mosquito, the best efforts of a government to wipe out the viral carriers can never achieve complete success.

Chua, as he listened to the drone of the engines and sat with little to do but reflect on the two frenetic weeks just past, was now, in mid-March, all but convinced that this was not an epidemic of Japanese encephalitis. He had grown the virus that was in the blood and spinal fluid of three recent patients. He had captured the culprit, and it was definitely nothing that he or Lam had ever seen before.

On March 1, 1999, Chua's lab at the University Hospital in Kuala Lumpur received those first three samples of bodily fluids and brain tissue—one from a truck driver in Sungai Nipah and two from victims in Bukit Pelanduk—just as inhabitants of those towns began collapsing in its emergency room. It was the doctors' first confirmation that the disease had gained a foothold in Negri Sembilan. "This rash of new patients was alarming," remembers Goh Khean Jin, a neurologist at the hospital.

The clinicians were disturbed by more than just the size of the outbreak. The symptoms fit the profile of Japanese encephalitis, but the victims did not. Because it is spread by insects, Goh explains, "JE usually affects the very young and the very old, and it strikes in somewhat random fashion. But we were seeing mostly adult males falling ill, and no children. In some families four people would get sick, whereas the fifth would not. Plus about three quarters of the patients had been vaccinated against JE at least once. So we started asking more questions, and we learned that almost all the patients either owned a pig farm or worked on one."

It began to look less and less like an insect-borne illness. So far everyone who caught it had touched a pig at some point—many while caring for animals that were coughing and wheezing with some strange sickness. That, too, was odd, because the JE virus does not harm a hog, its natural host.

Chua and Lam had been asked only to confirm that their specimens did indeed contain antibodies to JE, signaling that those patients had either been infected with the JE virus or vaccinated against it. But the two men decided to go a step further. If they could grow enough of the virus, they might get a look at it. "We thought it might be a mutant strain, one that the vaccine did not protect against," Lam says.

Chua dismissed his technicians, locked himself and one assistant in the biohazard lab, and began placing droplets of infected fluid onto cultures of kidney cells from pigs and monkeys. "He even injected them into mosquito larvae and suckling mice," Lam says. "We didn't really know what we were looking for. We just tried to cover the field."

Two or three times every day Chua checked the cells in the incubator for signs of infection. Many pathogens will grow only at a certain temperature or pH. Two years before, when Chau had isolated an enterovirus that sickened thousands in Malaysia, "it took 10 days to find the right conditions for growth," he recalls.

This agent was decidedly more aggressive. "It practically grew by itself—and very quickly," Chua says. Within three days the monkey cells began dying. By the fifth day many of those that were left had merged like water droplets into giant cellular blobs with multiple nuclei. The insect larvae—the host most susceptible to the JE virus—remained healthy.

For a week, Chua ran battery after battery of antibody tests on the viral isolate. The test for JE came up negative; measles also. Herpes simplex, dengue virus, panenterovirus, cytomegalovirus, respiratory syncytial virus—they tested for anything that might cause encephalitis. "All came up negative," he says. "Under the electron microscope," Lam recounts, "the viral particles looked very large. That was a clue that it might be a paramyxovirus," perhaps a cousin of the pathogens behind measles, mumps, and some other highly contagious diseases.

"I remember going down to Chua's lab that day," Goh says. "He said, 'Look, we've got a new virus!' I was very frightened. We had been touching these patients, cleaning them. And Dr. Chua and others in the lab were growing quite large quantities with no protection. But scientifically it was extremely exciting."

If it was indeed a paramyxovirus causing the encephalitis, then Lam knew that it could not be carried by insects: fogging and JE vaccination would not work. But he lacked the equipment to be certain, and he was pressed for time as new patients continued to pour in. Lam decided to accept an offer of assistance from an old friend at the CDC in Fort Collins. "When we tried to ship the virus to the U.S., one courier company after another turned us down," Lam says. So the day after making his discovery, Chua gingerly packed the infected blood, spinal fluid and bits of human brain into an airtight metal capsule, placed it in dry ice and headed for the airport.

Seventy-two hours later CDC scientists, guided by Chua, reproduced his results but also failed to identify the virus. "We had anticipated that and had forwarded samples on to CDC headquarters in Atlanta," Lam says. After another long, uneasy flight, Chua arrived in Atlanta to learn that the virologists there already had some disturbing news. The virus, which Lam and Chua named Nipah after the village of the man from whom the isolate was grown, was completely new to medicine. But it shared about 82 percent of its DNA sequence with a virus called Hendra, which had killed 14 race-horses and their trainer in 1994 in Queensland, Australia. Hendra is spread by fruit bats. It so happens that fruit bats live in almost every part of Malaysia, and they are not known for halting their flight at national boundaries.

Brian Mahy, head of the CDC's division of viral diseases, sent Lam an e-mail with the news and with an offer to send a team of 10 experts, including two Australian veterinarians, to help with the investigation. "I took the message to the director general at the ministry of health," Lam remembers, "and he approved the idea on the spot."

By the second week of March, the wards were filling up with encephalitis patients," recalls Patrick Tan, who was helping to run the university hospital's intensive care unit. "It was a steady stream: one per day on average. That indicated a big pool of illness out there, but we could not imagine how big. Our worst estimates were being exceeded almost daily."

As the encephalitis patients grew sicker and larger in number, Tan scrambled to find more ventilators and nurses. Elective surgeries were postponed. "At the peak, we were operating very close to our bare minimum standards for medical care," he admits in a soft English accent, his lips drawn tight above a neat bow tie. Families, having abandoned their homes, crowded the corridors. Morale plummeted. "The mortality was very high: sometimes three deaths a day," Tan explains. "You would see a son die at seven in the morning and his father pass away at noon. We were clearly dealing with something unknown and very threatening."

The threat was more than medical. The demand for pork, consumed almost exclusively by the ethnic Chinese minority, had fallen by 90 percent, kicking out one of the few stable pillars in an economy still reeling from last year's currency crisis. "Between the farmers and slaughterers and meat sellers and lorry drivers, something like 10 percent of the Chinese population here have been affected," Lam says. Although in Malaysia the large populations of Hindu Indian and Muslim Malay had not risen against the Chinese, as had happened in neighboring Indonesia, some Western observers feared that the news of a lethal disease borne by pigs could stoke religious hostility.

The scientists hoped that identifying the virus would suggest ways of stopping it. The CDC had confirmed Lam's hunch that the pathogen was a member of the Paramyxoviridae family. "That told us that it is an RNA virus surrounded by a lipid [fatty] envelope, and therefore it is easy to inactivate with heat or detergents," Lam says. The word went out that cooked pork was safe to eat, but still no one wanted to buy it. Nor did publicizing the fact that pig farms could be disinfected with soapy sprays stop the (largely Muslim) army from bulldozing pig farms to rubble as well. Few drugs affect them. The microbiologists could suggest only one, ribavirin. "It is expensive," Tan says. "But we had nothing else to give these patients, so we tried it."

But having a paramyxovirus as an enemy furrowed brows. Many paramyxoviruses cause respiratory infections that spread in aerosol form, which makes them particularly dangerous. "In fact, we knew that the virus gives pigs a terrible cough—they call it a 'one-mile cough' because you can hear it from a mile away. That is probably how it is spreading among pigs," Lam explains. "Even in humans we have shown that the virus is in the urine and the gargle. But we do not know how infectious those secretions are. It's just like HIV; the virus may be in the saliva but not in enough quantity to transmit the infection."

Even as the outbreak reached its peak in late March, not a single doctor or nurse had caught the Nipah virus. That suggested that it does not move easily from human to human. But to be safe, the CDC tentatively assigned the pathogen to Biosafety Level 4. That elite group of the most lethal, contagious agents—including the Ebola, Marburg and Lassa viruses—can be safely handled only in "hot zone" labs. There are just a few in the world, and none are close to Malaysia. To stop the outbreak and save as many lives as possible, three questions were now paramount: What does the Nipah virus do to the body? What animals can transmit it? And what is its natural host, the species in which the virus thrives but does not kill?

At the level of gross anatomy, the spectrum of damage wrought by Nipah virus on human bodies can be seen in the encephalitis wards of University Hospital as the neurologists make their morning rounds. Two weeks after the outbreak peaked, the ward is still full with 16 patients. Many are in the same vegetative state as the farmer whose eyeballs roll unresponsively while C. T. Tan, the chief neurologist, lifts their lids and shouts, "Look here! Look here!" in Chinese.

Goh walks up to another Chinese man lying still in bed,

seemingly asleep, although his heart monitor shows a pulse rate of 130. A young woman stands beside him. She pats his shoulder. "Hello, Mr. Ching?" Tan says. "Hello?" There is no response, except slight grimaces when one of the interns taps the man's knees with a rubber mallet. The woman clasps her latex-gloved hands tightly and looks at the doctor with fear in her eyes. The entourage moves on.

"This one gives me the greatest hope," Goh says as we approach a middle-aged woman. "Two weeks ago she was comatose and suffered tremors and seizures. Now she can speak a little and almost walk on her own." But after Tan has her take his arm and make a few halting steps, she stops and sways. Her eyelids droop shut. Goh cannot predict how fully she will recover.

Nor can anyone know yet whether those who struggle back to health will retain it. At the next bed a 31-year-old farmhand from Bukit Pelanduk convulses quietly as nurses attend to him. "We treated this man with ribavirin, and after a week he was well enough to go home," Goh recalls. "But then he returned with new symptoms." He was recovering from those, too, when suddenly a blood vessel burst in his brain. His pillow and sheets are stained red and brown. "Now his prognosis is very guarded," the earnest doctor whispers, lowering his head.

This relapse worries Goh. A second Australian horse trainer who caught Hendra in 1994 felt fine for 13 months, then developed encephalitis and died within days. It is possible that the hundreds who survived Nipah infection may still be in danger. "These people will need to be followed for several years to come," Goh says.

Several stories below the ward, neuropathologist Wong Kum Thong straps on a mask and apron and walks over to brightly colored buckets in shelves against the wall of the post-mortem room. "We need to be sure that we have observed all the possible changes caused by this disease," the thin neuropathologist says as he pulls a heart from a yellow bucket. Slicing through a thick layer of fat, he cuts a thin section of aorta and hands it to an assistant for labeling. The smell of formalin pricks the nostrils with a sting.

"There is only a small random chance of seeing an important phenomenon in any given section, so you have to look at many, many slides," Wong says. He has moved on to the stomach of one of the Nipah virus victims. Organs from more than a dozen are kept here for study. "There is no drama involved," Wong continues, "just hours on the microscope examining and reexamining samples. It is very tedious work—like most of science."

And yet there are many small moments of discovery. Wong's slides have revealed that the Nipah virus attacks the cells that line the blood vessels in virtually every organ, from brain to lungs to kidneys. That is one way it disables its victims: by inflaming the brain and fouling its blood supply. But the virus can also infect neurons directly, stuffing them with viral particles until they burst. There is undoubtedly more to discover.

"Come look at this," Wong says when he has finished his sections. Out of a blue bucket he lifts a large object suspended by string in the formalin. The brain in his hands is blotched with brown splatters and lines. "There are lots of amorphous hemorrhages here," he observes. But that is not what has caught Wong's attention. He puts a gloved finger to several small black spots, pinhead-size circles that dot many parts of the surface of the brain. "I've never seen anything like these before."

"They are too large to be point failures of capillaries," muses George Paul, the hospital's forensic pathologist. "Why are they there, and are there more inside? We should photograph those."

Wong nods. "If we could just get one more autopsy, I could do an electron microscope study of the blood vessels and neurons in the brain. It would be very elegant to show the virus there, doing its damage. We still do not know how long the virus remains viable after death."

The next day Chua is working in the biohazard lab, preparing serum samples for testing. There is as yet no definitive blood test for Nipah infection, but a screen for Hendra antibodies seems to work well enough. Still, it cannot tell Chua what he really wants to know: whether the immune system also fights Nipah virus with T cells, its strongest weapon, and whether it can wipe out the infection or, as with HIV, only send it into temporary remission.

The lab doorbell rings; it is Wong. He calls Chua over and speaks rapidly. "We have an encephalitis case in which a post-mortem is very likely." The 31-year-old patient has just died. "I think it is very important to get live virus from the brain if possible. But I don't want to expose myself. How should I bring the brain up to you?" Here is the opportunity both had been waiting for: a chance to survey the brain after the immune system engaged the enemy, won a battle but then lost the war.

The findings from that autopsy have become only one more piece in the puzzle. With luck and probably many years of work, Lam and his colleagues around the world may be able to find a vaccine for this new Malaysian encephalitis. In the meantime, the veterinarians are still searching for the natural host of the Nipah virus and for clues to how far it has spread.

The Malaysian government's solution to the outbreak—dispatching soldiers in chemical warfare gear to kill all pigs within five kilometers (three miles) of an infected farm—may have panicked some. Farmers around Bukit Pelanduk, unwilling to wait for the army or fearful that their buildings would be razed, dug large pits, herded their pigs in and buried them alive. Others clubbed their swine to death with planks. But it was effective. In just three weeks, 900,000 head of hog were obliterated. Encephalitis cases began to drop.

And yet weeks later the houses in Sungai Nipah and Bukit Pelanduk are shuttered. Some still have laundry on the line and toys abandoned in the yard. A few stray dogs have the streets to themselves but are to be shot on sight. The demand for pork is still 80 percent below what it was. The president of the livestock farmers association has reportedly predicted that an industry that once brought in 1.5 billion Malaysian ringgit ($395 million) a year will take more than five years to recover.

That may prove optimistic. In May 1999 a nationwide testing program revealed that the Nipah virus has spread to pig farms in other states: Selangor, Johore, Malacca, Penang—virtually the entire western half of the country. One of the Australian veterinary experts recently announced that a quarter of certain species of fruit bats collected in some regions carried antibodies to the virus. No one yet knows, however, whether bats are the virus's natural host—and if they are, what that means for pig farming in southeast Asia.

"We can only hope that the disease is cyclical," Lam says, "and that there will be many years between each cycle," time enough to develop a vaccine or find an effective treatment. While we are hoping, Patrick Tan adds, let us hope there are no nastier viruses than Nipah about to break from their ecosystem to ours. "This experience will stand us in good stead if we have a further encounter with an outbreak," he says. "We learned that the crisis led to a unity of purpose; people were prepared to put themselves second."

And yet a highly skilled scientific team, aiding a government possessed of great power and the willingness to use it, has been humbled by the escape of its new foe. "We cannot be too proud," Tan warns. The virus next time may be even worse.

Whereas nature has long hurled pathogens at humankind, in the 21st century, that dynamic has taken a dangerous new twist. People now hurl lethal pathogens at people as an act of war.

The weapons of choice for these attacks include the most deadly germs known to man, including smallpox, pneumonic plague, and anthrax, all noted villains in civilization's past. Even more recently discovered exotic viruses, once isolated, can be stockpiled and employed as bombs of mass destruction in a biological warfare field. "Facing an Ill Wind," which follows, gives an overview of the current threats and how the government is responding.

ling bacteria is like hitti
get--just when you think y
m, they mutate on you. Thro
ruse and misuse of antibiot
elerated the development of
teria that survive conventi
c attack. drugs pot
medical community preach
antibiotics and teach pati

Facing an Ill Wind

Tim Beardsley

T he specter of mass civilian casualties resulting from an attack with biological weapons has long been a worst-case scenario mulled over by defense planners. But in recent years the threat has moved to the front of the U.S. policy agenda, driven by a series of unwelcome revelations. Soviet èmigrè Ken Alibek, former deputy head of the secret laboratory known as Biopreparat, has recounted how the former Soviet Union manufactured tens of tons of "weaponized" smallpox virus; which is highly contagious and would likely spread rapidly in the now largely unimmunized U.S. population. The Soviets also produced weapons based on pneumonic plague and anthrax, Alibek has charged, and they experimented with aerosolized Ebola and Marburg viruses, which cause massive hemorrhaging.

Disclosures about sophisticated anthrax-based biological weapons developed by Iraq have also contributed to growing apprehension, as did the discovery that the Aum Shinrikyo cult in Japan released anthrax spores and botulinum in Tokyo nine times before it carried out its deadly 1995 subway attack with the nerve gas sarin. The Aum's attempted germ attacks failed because the group's biologists cultured the strain of anthrax used to make vaccine, which is harmless; had they used a potent culture, the outcome might have been very different. (No one knows why the botulism attack failed.)

Others are less confident. Donald A. Henderson of Johns Hopkins University, who spearheaded the World Health Organization's successful campaign to eradicate smallpox, counters that widely known advances in fermentation and dispersion technology make it easier than ever for a malefactor to grow substantial quantities of some deadly agents and use them. Unlike nuclear or chemical weapons, biological weapons can be made with readily available materials or equipment. Many deadly agents, including plague and anthrax, can be found in nature. (Only two declared locations in the world hold the smallpox virus, but Henderson says he is "persuaded" that smallpox is being worked on at undeclared laboratories in Russia and possibly elsewhere.) Henderson believes 10 to 12 countries are now researching biological weapons. Moreover, thanks to domestic economic woes, Russian microbiologists are often targets for recruitment by foreign powers.

Advances in molecular biology could make engineering a superpathogen more feasible, according to Steven M. Block of Princeton University, the only molecular biologist on the panel of defense advisers known as the Jasons. Block says smallpox or anthrax engineered for extra lethality is "very credible indeed."

Most agents produce flulike symptoms in the early stages of infection, so the first victims would most likely be sent home with a diagnosis of a nonspecific viral syndrome. Only when authorities noticed unusual deaths would the alarm be raised. At that point, public demand for prophylactic medications would quickly become intense. Yet in 1999, there were only some seven million doses of smallpox vaccine in the U.S., and scaling up production would take at least 36 months, according to Henderson. He estimates that an attack with aerosolized smallpox virus that initially infected just 100 people would within a few weeks paralyze a large part of the country: by the time the first cases had been diagnosed, people would have carried the infection to other cities.

Dozens of different agents might conceivably be employed as a weapon. Indeed, the first successful biological attack in the U.S., which was not recognized as such at the time, was with salmonella. Followers of Bhagwan Shree Rajneesh put the bacteria in salad bars in restaurants in The Dalles, Ore., in 1984, sickening several hundred people. But Henderson says anthrax, smallpox and plague represent by far the greatest threats.

The administration has proposed steep budget increases to counter biological threats against civilians. Surveillance for odd outbreaks of disease is being stepped up by 22 percent, to $86 million, regional laboratories are being established, and funds are being sought for 25 new emergency metropolitan medical teams. Research on vaccines is being boosted by $30 million, and specialized medicines are being stockpiled. The Department of Energy is working on new and better sensors and is studying airflow patterns in cities and around subways.

One focus is an attempt to prevent the spread of deadly agents with water curtains and giant balloons that would block off tunnels. Sandia scientists have also developed a noncorrosive foam that neutralizes chemical agents and effectively kills spores of a bacterium similar to anthrax. Some of the most far-out research is being funded by rapidly growing programs at the Defense Advanced Research Projects Agency (DARPA), which is researching sensitive detection devices and countermeasures that would work against a wide spectrum of agents. Many pathogens employ similar molecular mechanisms in the early stages of infection, notes Shaun B. Jones, head of DARPA's Unconventional Pathogen Countermeasures program. Many, too, share similar mechanisms of damage. Those insights make a search for broad-spectrum agents worthwhile, Jones maintains.

DARPA is also funding projects in which red blood cells are modified. Microbiologists Mark Bitensky and Ronald Taylor have shown that enzymatic complexes and antibodies can be added to the surfaces of red blood cells that give them the ability to bind pathogens. The antibodies carry the pathogens to the liver to be destroyed, and, remarkably, the lifetime of the red blood cells in the body is not affected. Maxygen in Santa Clara, Calif., is using a technique called DNA shuffling, which randomly combines potentially useful gene fragments to evolve potential DNA vaccines. James R. Baker, Jr. is developing liposomes and dendritic polymers that are safe to apply to the skin yet dissolve pathogens.

Some critics, however, maintain that high-tech may not be the best answer. The government once approached biological weapons "from the standpoint of vulnerability assessment, not threat assessment," says Jonathan B. Tucker of the Monterey Institute of International Studies. What is needed, he believes, is "a much better understanding of what motivates a group to use these weapons" so that terrorists can be stopped before they strike. Block of Princeton likewise emphasizes the great importance of human intelligence. Civil libertarians, however, worry about giving the military any permanent counter-terrorist role in the homeland.

If covert operations face difficulties, perhaps overt ones would be easier. And Barbara Hatch Rosenberg of the Federation of American Scientists argues that the U.S. could participate more constructively in the negotiations under way in Geneva aimed at strengthening the 1972 Biological and Toxin Weapons Convention. Senior officials "say the right things" about the convention, Rosenberg indicates. But she charges that the U.S. has repeatedly objected to a proposed inspection regime that would give it teeth, on the grounds that surprise visits by international inspectors might imperil commercial secrets—or compromise national security.

Some feel, moreover, that scientists themselves could do more to oppose biological terrorism. Just as physicists became active in the movement to prevent nuclear war in the past century, Block notes, "I would hope and expect biological scientists will take a leading role in anti-biological weapons activity."

Confronted with the reality of bioterrorism, science has responded with ideas for high-tech protection. One possible candidate for detection are biochips, a kind of molecular flypaper that binds to target pathogens and lights up if anything is detected. High-tech innovation like the one described in the following article may help quickly identify pathogens unleashed as an act of war or of terrorism.

Bioagent Chip

David Pescovitz

Speed is of the essence in successfully containing a biological warfare attack. Quickly identifying the agent and how to treat those who have been exposed are keys to controlling an outbreak and minimizing its destructiveness. A handheld device containing a laboratory-on-a-chip may just be the answer. The result of breakthroughs in biology, chemistry and micromanufacturing, the instrument can immediately alert investigators to even the slightest hint of anthrax or smallpox in the air.

Although there are myriad proposals for building these biosensors, the double whammy of identifying a particular bioagent in less than two minutes, and doing so given a sample of only a few cells, has been difficult to achieve. "There are many diseases that are as effective as influenza—they can affect you at the single- or a few-particle level," says Mark A. Hollis, manager of the biosensor technologies group at the Massachusetts Institute of Technology Lincoln Laboratory, where a collaborative effort with M.I.T. biologist Jianzhu Chen and his colleagues hopes to deliver a prototype biosensor in less than 18 months. The work is part of the

Defense Advanced Research Projects Agency's four-year, $24-million Tissue Based Biosensors program, which funds research by about a dozen universities and private firms.

Mouse B cells power the device. Part of the immune system, B cells express antibodies on their surfaces that bind to particular infectious particles. For example, most humans harbor B cells for pathogens that cause colds, polio, tetanus and other diseases. When a B cell binds to the intruder that it is built to recognize, a biochemical cascade occurs in the cell, triggering the body's immune system to rally to the defense. "We're leveraging off probably 600 to 800 million years of genetic engineering that nature has already done to recognize an infectious agent," Hollis observes.

With the design legwork out of the way courtesy of basic biology, Hollis's colleagues genetically engineer the B cells to respond to particular biowarfare agents. To know that the B cells have actually gone into action, the researchers plug into B cells another gene from a jellyfish called Aequorea. This gene enables the jellyfish to glow with the bioluminescent protein aequorin. The aequorin instantly emits light when triggered by calcium ions—a substance that is produced when the bio-agent-induced cascade occurs in the B cell. The entire process, from detection to bioluminescence, takes less than a second, beating any human handiwork to date.

Other methods have matched either the speed or the sensitivity of the B cells, but not both. The record for analyses using the polymerase chain reaction of a bioagent, Hollis says, is about 12 minutes, based on a pristine sample containing more than 20 organisms. Immunoassay techniques, which also use an antibody-capture methodology, are approaching the requisite speed but lack sensitivity: a sample containing at least several thousand copies of the organism is needed to identify an agent. In contrast, "only one infectious particle is sufficient to trigger a B cell because that's the way nature designed it," Hollis notes. "It's a beautifully sensitive system."

Currently the biosensor is a 25-millimeter-square plastic chip that has a meandering flow line running through it. One-to two-millimeter-square patches, containing 10,000 B cells engineered for an individual agent, line the surface of the channel. A strict diet combined with a room-temperature climate keeps the cells in their place by naturally discouraging cell division. Even hungry and cold, they stick to the task at hand.

Elegant microfluidics, also developed at Lincoln, direct the sample and nutrient media through the channel, where a charge-coupled device (CCD) like those found in camcorders detects even a single B cell firing. Identification based on five to 10 particles per sample has been demonstrated, and Hollis expects no problems detecting deadly bioagent particles in even the smallest numbers.

The biosensor, too, is naturally robust: exhaust, dirt and other contaminants that make the working environment considerably less than hospitable, compared with a B cell's traditional home inside the body, don't trick the cells into misfiring. "There's a lot of stuff in your blood, and these things are designed not to respond to any of it other than the virus they're intended for," remarks Hollis, who points out that the same B-cell-based biosensing technology developed for military use could be employed for instant viral identification in a doctor's office.

The last big question on Hollis's research agenda—whether the cells will reset after having fired—may not even matter in the group's latest vision for a handheld biosensor: a proposed optical-electronic box would read the photons emitted by a swappable and disposable biosensor chip, which would cost just a few dollars. "If you are hit with a biological attack," Hollis says, "you'll probably want to take the chip out and send it off to Washington for confirmation." Probably so.

The extremes to which people will go to kill other people are evident in the horrors of biological warfare. And yet, it is not a stratagem wholly new to our century—the Mongols employed it in the 14th century by catapulting plague-infested corpses over the walls of the Russian city of Kaffa. But in the 21st century, whole nations could be at risk of quick death at the hands of war pathogens. Worse, once unleashed, many of these agents of death could sweep indiscriminately around the globe.

The following piece by Leonard Cole was published five years before the September 11th attack on the United States and the wave of biological warfare that followed but still rings true.

The Specter of
Biological Weapons

Leonard A. Cole

I n 1995, on a whim, I asked a friend: Which would worry you more, being attacked with a biological weapon or a chemical weapon? He looked quizzical. "Frankly, I'm afraid of Alzheimer's," he replied, and we shared a laugh. He had elegantly dismissed my question as an irrelevancy. In civilized society, people do not think about such things.

The next day, on March 20, the nerve agent sarin was unleashed in the Tokyo subway system, killing 12 people and injuring 5,500. In Japan, no less, one of the safest countries in the world. I called my friend, and we lingered over the coincidental timing of my question. A seemingly frivolous speculation one day, a deadly serious matter the next.

That thousands did not die from the Tokyo attack was attributed to an impure mixture of the agent. A tiny drop of sarin, a chemical weapon originally developed in Germany in the 1930s, can kill within minutes after skin contact or inhalation of its vapor. Like all other nerve agents, sarin blocks the action of acetylcholinesterase, an enzyme necessary for the transmission of nerve impulses.

The cult responsible for the sarin attack, Aum Shinrikyo ("Supreme Truth"), was developing biological agents as well. If a chemical attack is frightening, a biological weapon poses a worse nightmare. Chemical agents are inanimate, but bacteria, viruses and other live agents may be contagious and reproductive. If they become established in the environment, they may multiply. Unlike any other weapon, they can become more dangerous over time.

Certain biological agents incapacitate, whereas others kill. The Ebola virus, for example, kills as many as 90 percent of its victims in little more than a week. Connective tissue liquefies; every orifice bleeds. In the final stages, Ebola victims become convulsive, splashing contaminated blood around them as they twitch, shake and thrash to their deaths.

For Ebola, there is no cure, no treatment. Even the manner in which it spreads is unclear, by close contact with victims and their blood, bodily fluids or remains or by just breathing the surrounding air. Recent outbreaks in Zaire prompted the quarantine of sections of the country until the disease had run its course.

The horror is only magnified by the thought that individuals and nations would consider attacking others with such viruses. In October 1992 Shoko Asahara, head of the Aum Shinrikyo cult, and 40 followers traveled to Zaire, ostensibly to help treat Ebola victims. But the group's real intention, according to an October 31, 1995, report by the U.S. Senate's Permanent Subcommittee on Investigations, was probably to obtain virus samples, culture them and use them in biological attacks.

Interest in acquiring killer organisms for sinister purposes is not limited to groups outside the U.S. On May 5, 1995, six weeks after the Tokyo subway incident, Larry Harris, a laboratory technician in Ohio, ordered the bacterium that causes bubonic plague from a Maryland biomedical supply firm. The company, the American Type Culture Collection in Rockville, Md., mailed him three vials of *Yersinia pestis*.

Harris drew suspicion only when he called the firm four days after placing his order to find out why it had not arrived. Company officials wondered about his impatience and his apparent unfamiliarity with laboratory techniques, so they contacted federal authorities. He was later found to be a member of a white supremacist organization. In November 1995 he pled guilty in federal court to mail fraud.

To get the plague bacteria, Harris needed no more than a credit card and a false letterhead. Partially in response to this incident, an antiterrorism law enacted in April 1996 required the Centers for Disease Control and Prevention to monitor more closely shipments of infectious agents.

What would Harris have done with the bacteria? He claimed he wanted to conduct research to counteract Iraqi rats carrying "supergerms." But if he had cared to grow a biological arsenal, the task would have been frighteningly simple. By dividing every 20 minutes, a single bacterium gives rise to more than a billion copies in 10 hours. A small vial of microorganisms can yield a huge number in less than a week. For some diseases, such as anthrax, inhaling a few thousand bacteria—which would cover an area smaller than the period at the end of this sentence—can be fatal.

Kathleen C. Bailey, a former assistant director of the U.S. Arms Control and Disarmament Agency, has visited several biotechnology and pharmaceutical firms. She is "absolutely convinced" that a major biological arsenal could be built with $10,000 worth of equipment in a room 15 feet by 15. After all, one can cultivate trillions of bacteria at relatively little risk to one's self with gear no more sophisticated than a beer fermenter and a protein-based culture, a gas mask and a plastic overgarment.

Fortunately, biological terrorism has thus far been limited to very few cases. One incident occurred in September 1984, when about 750 people became sick after eating in restaurants in an Oregon town called The Dalles. In 1986 Ma Anand

Sheela confessed at a federal trial that she and other members of a nearby cult that had clashed with local Oregonians had spread salmonella bacteria on salad bars in four restaurants; the bacteria had been grown in laboratories on the cult's ranch. After serving two and a half years in prison, Sheela, who had been the chief of staff for the cult leader, Bhagwan Shree Rajneesh, was released and deported to Europe.

But as a 1992 report by the Office of Technology Assessment indicated, both biological and chemical terrorism have been rare. Also rare has been the use of biological agents as weapons of war. Perhaps the first recorded incident occurred in the 14th century, when an army besieging Kaffa, a seaport on the Black Sea in the Crimea in Russia, catapulted plague-infected cadavers over the city walls. In colonial America a British officer reportedly gave germ-infested blankets from a smallpox infirmary to Indians in order to start an epidemic among the tribes. The only confirmed instance in this century was Japan's use of plague and other bacteria against China in the 1930s and 1940s.

As the 20th century draws to a close, however, an unpleasant paradox has emerged. More states than ever are signing international agreements to eliminate chemical and biological arms. Yet more are also suspected of developing these weapons despite the treaties. In 1980 only one country, the Soviet Union, had been named by the U.S. for violating the 1972 Biological Weapons Convention, a treaty that prohibits the development or possession of biological weapons.

Since then, the number has ballooned. In 1989 Central Intelligence Agency director William Webster reported that "at least 10 countries" were developing biological weapons. By 1995, 17 countries had been named as biological weapons suspects, according to sources cited by the Office of Technology Assessment and at U.S. Senate committee hearings. They include Iran, Iraq, Libya, Syria, North Korea, Taiwan, Israel, Egypt, Vietnam, Laos, Cuba, Bulgaria, India, South Korea,

South Africa, China and Russia. (Russian leaders insist that they have terminated their biological program, but U.S. officials doubt that claim.)

The first five of these countries—Iran, Iraq, Libya, Syria and North Korea—are especially worrisome in view of their histories of militant behavior. Iraq, for example, has acknowledged the claims of U.N. inspectors that during the 1991 Persian Gulf War it possessed Scud missiles tipped with biological warheads. A 1994 Pentagon report to Congress cited instability in eastern Europe, the Middle East and Southwest Asia as likely to encourage even more nations to develop biological and chemical arms.

Reversing this trend should be of paramount concern to the community of nations. Indeed, the elimination of biological as well as chemical weaponry is a worthy, if difficult, goal. The failure of this effort may increase the likelihood of the development of a man-made plague from Ebola or some other gruesome agent.

Dedication to biological disarmament in particular should be enhanced by another grim truth: in many scenarios, a large population cannot be protected against a biological attack. Vaccines can prevent some diseases, but unless the causative agent is known in advance, such a safeguard may be worthless. Antibiotics are effective against specific bacteria or classes of biological agents, but not against all. Moreover, the incidence of infectious disease around the world has been rising from newly resistant strains of bacteria that defy treatment. In this era of biotechnology, especially, novel organisms can be engineered against which vaccines or antibiotics are useless.

Nor do physical barriers against infection offer great comfort. Fortunately, most biological agents have no effect on or through intact skin, so respiratory masks and clothing would provide adequate protection for most people. After a short while, the danger could recede as sunlight and ambient temperatures destroyed the agents. But certain microorganisms can persist indefinitely in an environment. Gruinard Island, off the coast of Scotland, remained infected with anthrax spores for 40 years after biological warfare tests were carried out there in the 1940s. And in 1981 Rex Watson, then head of Britain's Chemical and Biological Defense Establishment, asserted that if Berlin had been bombarded with anthrax bacteria during World War II, the city would still be contaminated.

Although many Israelis did become accustomed to wearing gas masks during the 1991 Persian Gulf War, it seems unrealistic to expect large populations of civilians to wear such gear for months or years, especially in warm regions. U.N. inspectors in Iraq report that in hot weather they can scarcely tolerate wearing a mask for more than 15 minutes at a time.

Calls for more robust biological defense programs have grown, particularly after the Persian Gulf War. Proponents of increased funding for biological defense research often imply that vaccines and special gear developed through such work can protect the public as well as troops. But the same truths hold for both the military and civilians: unless an attack organism is known in advance and is vulnerable to medical interventions, defense can be illusory.

Indeed, the Gulf War experience was in certain respects misleading. Iraq's biological weapons were understood to be anthrax bacilli and botulinum toxin. (Although toxins are inanimate products of microorganisms, they are treated as biological agents under the terms of the 1972 Biological Weapons Convention.) Both are susceptible to existing vaccines and treatments, and protection of military forces therefore seemed possible. Research that would lead to enhanced defense against these agents is thus generally warranted.

But the improbabilities of warding off attacks from less traditional agents deserve full appreciation. Anticipating that research can come up with defenses against attack organisms whose nature is not known in advance seems fanciful. Moreover, even with all its limitations, the cost of building a national civil defense system against biological and chemical weapons would be substantial. A 1969 United Nations report indicated that the expense of stockpiling gas masks, antibiotics, vaccines and other defensive measures for civilians could exceed $20 billion. That figure, when adjusted for inflation, would now be about $80 billion.

Vaccines and protective gear are not the only challenges to biological defense. Identifying an organism quickly in a battlefield situation, too, is problematic. Even determining whether a biological attack has been launched can be uncertain. Consequently, the Pentagon has begun to focus more on detection.

In May 1994 Deputy Secretary of Defense John Deutch produced an interagency report on counterproliferation activities concerning weapons of mass destruction. Biological agent detectors in particular, he wrote, were "not being pursued adequately." To the annual $110 million budgeted for the development of biological and chemical weapons detection, the report recommended adding $75 million. Already under way were Pentagon-sponsored programs involving such technologies as ion-trap mass spectrometry and laser-induced breakdown spectroscopy, approaches that look for characteristic chemical signatures of dangerous agents in the air. The army's hope, which its spokespersons admit is a long way from being realized, is to find a "generic" detector that can identify classes of pathogens.

Meanwhile the military is also advancing a more limited approach that identifies specific agents through antibody-antigen combinations. The Biological Integrated Detection System (BIDS) exposes suspected air samples to antibodies that react with a particular biological agent. A reaction of the antibody would signify the agent is present, a process that takes about 30 minutes.

BIDS can now identify four agents through antibody-antigen reactions: *Bacillus anthracis* (anthrax bacterium), *Y. pestis* (bubonic plague), botulinum toxin (the poison released by botulism organisms) and *staphylococcus enterotoxin B* (released by certain staph bacteria). Laboratory investigations to identify additional agents through antibody-antigen reactions are in progress. But scores of organisms and toxins are viewed as potential warfare agents. Whether the full range, or even most, will be detectable by BIDS remains uncertain.

The most effective safeguard against biological warfare and biological terrorism is, and will be, prevention. To this end, enhanced intelligence and regulation of commercial orders for pathogens are important. Both approaches have been strengthened by provisions in the antiterrorism bill enacted in 1996. At the same time, attempts to identify and control emerging diseases are gaining attention. One such effort is ProMED (Program to Monitor Emerging Diseases), which was proposed in 1993 by the 3,000-member Federation of American Scientists.

Although focusing on disease outbreaks in general, supporters of ProMED are sensitive to the possibility of man-made epidemics. The ProMED surveillance system would include developing baseline data on endemic diseases throughout the world, rapid reporting of unusual outbreaks, and responses aimed at containing disease, such as providing advice on trade and travel. Such a program could probably distinguish disease outbreaks from hostile sources more effectively than is currently possible.

In addition, steps to strengthen the 1972 Biological Weapons Convention through verification arrangements—including on-site inspections—should be encouraged. The 139 countries that are parties to the convention are expected to discuss incorporating verification measures at a review conference in December of this year. After the last review conference, in 1991, a committee to explore such measures was established. VEREX, as the group was called, has listed various possibilities ranging from surveillance of the scientific literature to on-site inspections of potential production areas, such as laboratories, breweries and pharmaceutical companies.

Given the ease with which bioweapons can be produced, individuals will always be able to circumvent international agreements. But the absence of such agents from national arsenals—and tightened regulations on the acquisition and transfer of pathogens—will make them more difficult to obtain for hostile purposes. Verification can never be foolproof, and therefore some critics argue that verification efforts are a waste of time. Proponents nonetheless assert that sanctions following a detected violation would provide at least some disincentive to cheaters and are thus preferable to no sanctions at all. Furthermore, a strengthened global treaty underscores a commitment by the nations of the world not to traffic in these weapons.

The infrequent use of biological weapons to date might be explained in many ways. Some potential users have probably lacked familiarity with how to develop pathogens as weapons; moreover, they may have been afraid of infecting themselves. Nations and terrorists alike might furthermore be disinclined to use bioagents because they are by nature unpredictable. Through mutations, a bacterium or virus can gain or lose virulence over time, which may be contrary to the strategic desires of the people who released it. And once introduced into the environment, a pathogen may pose a threat to anybody who goes there, making it difficult to occupy territory.

But beneath all these pragmatic concerns lies another dimension that deserves more emphasis than it generally receives: the moral repugnance of these weapons. Their ability to cause great suffering, coupled with their indiscriminate character, no doubt contributes to the deep-seated aversion most people have for them. And that aversion seems central to explaining why bioweapons have so rarely been used in the past. Contrary to analyses that commonly ignore or belittle the phenomenon, this natural antipathy should be appreciated and exploited. Even some terrorists could be reluctant to use a weapon so fearsome that it would permanently alienate the public from their cause.

In recognition of these sentiments, the 1972 Biological Weapons Convention describes germ weaponry as "repugnant to the conscience of mankind." Such descriptions have roots that reach back thousands of years. (Not until the 19th century were microorganisms understood to be the cause of infection; before then, poison and disease were commonly seen as the same. Indeed, the Latin word for "poison" is "virus.")

Among prohibitions in many civilizations were the poisoning of food and wells and the use of poison weapons. The Greeks and Romans condemned the use of poison in war as a violation of ius gentium—the law of nations. Poisons and other weapons considered inhumane were forbidden by the Manu Law of India around 500 B.C. and among the Saracens 1,000 years later. The prohibitions were reiterated by Dutch statesman Hugo Grotius in his 1625 opus The Law of War and Peace, and they were, for the most part, maintained during the harsh European religious conflicts of the time.

Like the taboos against incest, cannibalism and other widely reviled acts, the taboo against poison weapons was sometimes violated. But the frequency of such violations may have been minimized because of their castigation as a "defalcation of proper principles," in the words of the 18th- and 19th-century English jurist Robert P. Ward. Under the law of nations, Ward wrote, "Nothing is more expressly forbidden than the use of poisoned arms."

Historian John Ellis van Courtland Moon contends that growing nationalism in the 18th century weakened the disinclinations about poison weapons. As a result of what Moon calls "the nationalization of ethics," military necessity began to displace moral considerations in state policies; nations were more likely to employ any means possible to attain their aims in warfare.

In the mid-19th century, a few military leaders proposed that toxic weapons be employed, although none actually were. Nevertheless, gas was used in World War I. The experience of large-scale chemical warfare was so horrifying that it led to the 1925 Geneva Protocol, which forbids the use of chemical and bacteriological agents in war. Images of victims gasping, frothing and choking to death had a profound impact. The text of the protocol reflects the global sense of abhorrence. It affirmed that these weapons had been "justly condemned by the general opinion of the civilized world."

Chemical and biological weapons were used in almost none of the hundreds of wars and skirmishes in subsequent decades—until Iraq's extensive chemical attacks during the Iran-Iraq war. Regrettably, the international response to Iraqi behavior was muted or ineffective. From 1983 until the war ended in 1988, Iraq was permitted to get away with chemical murder. Fear of an Iranian victory stifled serious outcries against a form of weaponry that had been universally condemned.

The consequences of silence about Iraq's behavior, though unfortunate, were not surprising. Iraqi ability to use chemical weapons with impunity, and their apparent effectiveness against Iran, prompted more countries to arm themselves with chemical and biological weapons. Ironically, in 1991 many of the countries that had been silent about the Iraqi chemical attacks had to face a chemically and biologically equipped Iraq on the battlefield.

To its credit, since the Persian Gulf War, much of the international community has pressed Iraq about its unconventional weapons programs by maintaining sanctions through the U.N. Security Council. Council resolutions require elimination of Iraq's biological weapons (and other weapons of mass destruction), as well as information about past programs to develop them. Iraq has been only partially forthcoming, and U.N. inspectors continue to seek full disclosure.

But even now, U.N. reports are commonly dry recitations. Expressions of outrage are rare. Any country or group that develops these weapons deserves forceful condemnation. We need continuing reminders that civilized people do not traffic in, or use, such weaponry. The agreement by the U.S. and Russia to destroy their chemical stockpiles within a decade should help.

Words of outrage alone, obviously, are not enough. Intelligence is important, as are controls over domestic and international shipments of pathogens and enhanced global surveillance of disease outbreaks. Moreover, institutions that reinforce positive behavior and values are essential.

The highest priority in this regard is implementation of the Chemical Weapons Convention, which outlaws the possession of chemical weapons. It lists chemicals that signatory nations must declare to have in their possession. Unlike the Biological Weapons Convention, the chemical treaty has extensive provisions to verify compliance, including short-notice inspections of suspected violations. It also provides added inducements to join through information exchanges and commercial privileges among the signatories.

In 1993 the chemical treaty was opened for signature. By October 1996, the pact had been signed by 160 countries and ratified by 64, one less than the number required for the agreement to enter into force. One disappointing holdout is the U.S. In part because of disagreements over the treaty's verification provisions, the U.S. Senate recently delayed a vote on the pact.

Implementing this chemical weapons treaty should add momentum to the current negotiations over strengthening the Biological Weapons Convention. Conversely, failure of the Chemical Weapons Convention to fulfill expectations will dampen prospects for a verification regime for the biological treaty. The most likely consequence would be the continued proliferation of chemical and biological arsenals around the world. The longer these weapons persist, the more their sense of illegitimacy erodes, and the more likely they will be used— by armies and by terrorists.

As analysts have noted, subnational groups commonly use

the types of weapons that are in national arsenals. The absence of biological and chemical weapons from national military inventories may diminish their attractiveness to terrorists. According to terrorism expert Brian M. Jenkins, leaders of Aum Shinrikyo indicated that their interest in chemical weapons was inspired by Iraq's use of chemicals during its war with Iran.

Treaties, verification regimes, global surveillance, controlled exchanges of pathogens—all are the muscle of arms control. Their effectiveness ultimately depends on the moral backbone that supports them and the will to enforce them rigorously. By underscoring the moral sense behind the formal exclusion of biological weapons, sustaining their prohibition becomes more likely.

Defenses against Biological Weapons

Respirator or gas mask. Filters, usually made of activated charcoal, must block particles larger than one micron. Overgarments are also advisable to protect against contact with open wounds or otherwise broken skin.

Protective shelter. Best if a closed room, ideally insulated with plastic or some other nonpermeable material and ventilated with filtered air.

Decontamination. Such traditional disinfectants as formaldehyde are effective for sterilizing surfaces.

Vaccination. Must be for specific agent. Some agents require several inoculations over an extended period before immunity is conferred. For many agents, no vaccine is available.

Antibiotics. Effective against some but not all bacterial agents (and not effective against viruses). For some susceptible bacteria, antibiotic therapy must begin within a few hours of exposure—before symptoms appear.

Detection systems. Only rudimentary field units currently available for a few specific agents. Research is under way to expand the number of agents that can be detected in battlefield situations or elsewhere.

Potential Biological Agents

Bacillus anthracis. Causes anthrax. If bacteria are inhaled, symptoms may develop in two to three days. Initial symptoms resembling common respiratory infection are followed by high fever, vomiting, joint ache and labored breathing, and internal and external bleeding lesions. Exposure may be fatal. Vaccine and antibiotics provide protection unless exposure is very high.

Botulinum toxin. Cause of botulism, produced by Clostridium botulinum bacteria. Symptoms appear 12 to 72 hours after ingestion or inhalation. Initial symptoms are nausea and diarrhea, followed by weakness, dizziness and respiratory paralysis, often leading to death. Antitoxin can sometimes arrest the process.

Yersinia pestis. Causes bubonic plague, the Black Death of the Middle Ages. If bacteria reach the lungs, symptoms–including fever and delirium–may appear in three or four days. Untreated cases are nearly always fatal. Vaccines can offer immunity, and antibiotics are usually effective if administered promptly.

Ebola virus. Highly contagious and lethal. May not be desirable as a biological agent because of uncertain stability outside of animal host. Symptoms, appearing two or three days after exposure, include high fever, delirium, severe joint pain, bleeding from body orifices, and convulsions, followed by death. No known treatment.

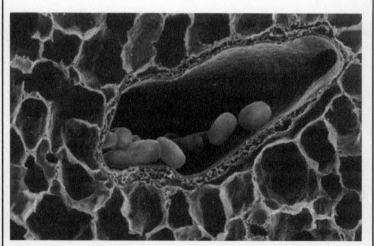

Photocomposite of *Bacillus anthracis* in lung bronchiole

.ling bacteria is like hitt
.get--just when you think y
.m, they mutate on you. Thr
.ruse and misuse of antibio
.farme... ...ttingly
.elerated the development o
.teria that survive convent
.c attack. To keep drugs po
. medical community preach
. antibiotics and teach pati

Conclusion

It may be difficult to fathom how an area of human curiosity—the pursuit of science—could give rise to bioagents capable of being used as weapons of mass destruction. Yet the science that gave rise to biological warfare is not to blame, any more than a copper wire in the Hiroshima nuclear bomb is to blame for the world's first nuclear attack. Our science is a tool, and as such it mirrors its makers and serves our intent.

Whether its revelations are used for war or used for peace, the study of microscopic life is a science full of hope and full of wonder. Could the human family learn something about it evolution and genetic make up from the humble bacteria? Or will we engineer a virus to expand our neural pathways or to fight pathogens? At the beginning of the 21st century, this science, and all our sciences, stand at the threshold of incredible discoveries.

And, as always, humankind's scientific and technical achievements will reflect our highest hopes and our deepest fears.

lling bacteria is like hitt
rget--just when you think y
em, they mutate on you. Thr
eruse and misuse of antibio
d farmers ha   wittingly
celerated the development o
teria that survive convent
ic attack. To keep drugs po
e medical community preach
antibiotics and teach pati

Index